高等教育安全科学与工程类系列规划教材
高等院校安全工程类特色专业系列规划教材

安全工程专业英语

主　编　黄志安　张英华
副主编　高玉坤　王　辉
参　编　孙传武　张　歌　燕立凯
　　　　王树祎　杨　锐
主　审　金龙哲　王　凯

机 械 工 业 出 版 社

本书共20个单元，其中的科技英语文章均按安全科学与工程学科所涉及的领域选取，涵盖施工安全、食品安全、矿井安全、特种设备安全、安全评价与风险管理、应急救援等多个方面。本书每个单元主要由课文和相应的阅读材料组成，其后都附有生词和短语注解，还编入了练习题，方便读者理解与学习。每个单元还单独设置一个模块，介绍了科技英语的阅读方法和技巧、科技英语的语言特点，以及科技英语的翻译技巧，同时列举了大量的例句，有助于理解与掌握；对科技英语写作要点进行了简单介绍，并列举了科技英语写作中的常用句型。

本书内容广泛，形式新颖，主要作为高等院校安全科学与工程类专业英语本科教材，也可作为安全工程技术和管理人员的学习参考用书。

图书在版编目（CIP）数据

安全工程专业英语/黄志安，张英华主编．—北京：机械工业出版社，2018.4（2024.7重印）

高等教育安全科学与工程类系列规划教材　高等院校安全工程类特色专业系列规划教材

ISBN 978-7-111-59276-1

Ⅰ.①安…　Ⅱ.①黄…②张…　Ⅲ.①安全工程－英语－高等学校－教材　Ⅳ.①X93

中国版本图书馆CIP数据核字（2018）第036144号

机械工业出版社（北京市百万庄大街22号　邮政编码100037）

策划编辑：冷　彬　责任编辑：冷　彬　杨　洋　马军平

责任校对：李　伟　封面设计：张　静

责任印制：李　昂

北京捷迅佳彩印刷有限公司印刷

2024年7月第1版第3次印刷

184mm×260mm·15.5印张·416千字

标准书号：ISBN 978-7-111-59276-1

定价：38.00元

安全科学与工程类专业教材编审委员会

序

“安全工程”本科专业是在1958年建立的“工业安全技术”“工业卫生技术”和1983年建立的“矿山通风与安全”本科专业基础上发展起来的。1984年，国家教委将“安全工程”专业作为试办专业列入普通高等学校本科专业目录之中。1998年7月6日，教育部发文颁布了《普通高等学校本科专业目录》，“安全工程”本科专业（代号：081002）属于工学门类的“环境与安全类”（代号：0810）学科下的两个专业之一[㊀]。1958～1996年年底，全国各高校累计培养安全工程专业本科生8130人。到2005年年底，在教育部备案的设有安全工程本科专业的高校已达75所，2005年全国安全工程专业本科招生人数近3900名[㊁]。

按照《普通高等学校本科专业目录》的要求，以及院校招生和专业发展的需要，原来已设有与“安全工程”专业相近但专业名称有所差异的高校，现也大都更名为“安全工程”专业。专业名称统一后的“安全工程”专业，专业覆盖面大大拓宽[㊀]。同时，随着经济社会发展对安全工程专业人才要求的更新，安全工程专业的内涵也发生了很大变化，相应的专业培养目标、培养要求、主干学科、主要课程、主要实践性教学环节等都有了不同程度的变化，学生毕业后的执业身份是注册安全工程师。但是，安全工程专业的教材建设与专业的发展出现尚不适应的新情况，无法满足和适应高等教育培养人才的需要。为此，组织编写、出版一套新的安全工程专业系列教材已成为众多院校的翘首之盼。

机械工业出版社是有着悠久历史的国家级优秀出版社，在高等学校安全工程学科教学指导委员会的指导和支持下，根据当前安全工程专业教育的发展现状，本着“大安全”的教育思想，进行了大量的调查研究工作，聘请了安全科学与工程领域一批学术造诣深厚、实践经验丰富的教授和专家，组织成立了“安全工程专业教材编审委员会”（以下简称“编审委”），决定组织编写“高等教育安全工程系列‘十一五’规划教材”[㊂]，并先后于2004年8月（衡阳）、2005年8月（葫芦岛）、2005年12月（北京）、2006年4月（福州）组织召开了一系列安全工程专业本科教材建设研讨会，就安全工程专业本科教育的课程体系、课程教学内容、教材建设等问题反复进行了研讨，在总结以往教学改革、教材编写经验的基础上，以推动安全工程专业教学改革和教材建设为宗旨，进行顶层设计，

㊀ 按《普通高等学校本科专业目录》（2012版），“安全工程”本科专业（专业代码：082901）属于工学学科的“安全科学与工程类”（专业代码：0829）下的专业。

㊁ 各高校安全工程本科每年的招生数量可以通过高等学校安全工程学科教学指导委员会主办的“全国高等院校安全工程学科教育数据和信息系统”查询（www. cosha. org. cn）。

㊂ 自2012年更名为“高等教育安全科学与工程类系列规划教材”。

制订总体规划、出版进度和编写原则，计划分期分批出版30余门课程的教材，以尽快满足全国众多院校的教学需要，以后再根据专业方向的需要逐步增补。

由安全学原理、安全系统工程、安全人机工程学、安全管理学等课程构成的学科基础平台课程，已被安全科学与工程领域学者认可并达成共识。本套系列教材编写、出版的基本思路是，在学科基础平台上，构建支撑安全工程专业的工程学原理与由关键性的主体技术组成的专业技术平台课程体系，编写和出版系列教材来支撑这个体系。

本套系列教材体系设计的原则是，重基本理论，重学科发展，理论联系实际，结合学生现状，体现人才培养要求。为保证教材的编写质量，本着“主编负责，主审把关”的原则，编审委组织专家分别对各门课程教材的编写大纲进行认真仔细的评审。教材初稿完成后又组织同行专家对书稿进行研讨，编者数易其稿，经反复推敲定稿后才最终进入出版流程。

作为一套全新的安全工程专业系列教材，其“新”主要体现在以下几点：

体系新。本系列教材从“大安全”的专业要求出发，从整体上考虑、构建支撑安全工程学科专业技术平台的课程体系和各门课程的内容安排，按照教学改革方向要求的学时，统一协调与整合，形成一个完整的、各门课程之间有机联系的系列教材体系。

内容新。本系列教材的突出特点是内容体系上的创新。它既注重知识的系统性、完整性，又特别注意各门学科基础平台课之间的关联，更注意后续的各门专业技术课与先修的学科基础平台课的衔接，充分考虑了安全工程学科知识体系的连贯性和各门课程教材间知识点的衔接、交叉和融合问题，努力消除相互关联课程中内容重复的现象，突出安全工程学科的工程学原理与关键性的主体技术，有利于学生的知识和技能的发展，有利于教学改革。

知识新。本套系列教材的主编大多由长期从事安全工程专业本科教学的教授担任，他们一直处于教学和科研的第一线，学术造诣深厚，教学经验丰富。在编写教材时，他们十分重视理论联系实际，注重引入新理论、新知识、新技术、新方法、新材料、新装备、新法规等理论研究、工程技术实践成果和各校教学改革的阶段性成果，充实与更新了知识点，增加了部分学科前沿方面的内容，充分体现了教材的先进性和前瞻性，以适应时代对安全工程高级专业技术人才的培育要求。本套系列教材中凡涉及安全生产的法律法规、技术标准、行业规范，全部采用最新颁布的版本。

安全是人类最重要和最基本的需求，是人民生命与健康的基本保障。一切生活、生产活动都源于生命的存在。如果人们失去了生命，一切都无从谈起。全世界平均每天发生约68.5万起事故，造成约2200人死亡的事实，使我们确认，安全不是别的什么，安全就是生命。安全生产是社会文明和进步的重要标志，是经济社会发展的综合反映，是落实以人为本的科学发展观的重要实践，是构建和谐社会的有力保障，是全面建成小康社会、统筹经济社会全面发展的重要内容，是实施可持续发展战略的组成部分，是各级政府履行市场监管和社会管理职能的基本任务，是企业生存、发展的基本要求。国内外实践证明，安全生产具有全局性、社会性、长期性、复杂性、科学性和规律性的特点，随着社会的不断进步，工业化进程的加快，安全生产工作的内涵发生了重大变化，它突破了时间和空间的限制，存在于人们日常生活和生产活动的全过程中，成为一个复杂多变的社会问题在安全领域的集中反映。安全问题不仅对生命个体非常重要，而且对社会稳定和经济发展产生重要影响。党的十六届五中全会提出“安全发展”的重要战略理念。安全发展是科学发展观理论体系的重要组成部分，安全发展与构建和谐社会有着密切的内在联系，以人为本，首先就是要以人的生命为本。“安全·生命·稳定·发展”是一个良性循环。安全科技工作者在促进、保证这一良性循环中起着重要作用。安全科技人才匮乏是我国安全生产形势严峻的重要原因之一。加快培养安全科技人才也是解开安全难题的钥匙之一。

高等院校安全工程专业是培养现代安全科学技术人才的基地。我深信，本套系列教材的出版，将对我国安全工程本科教育的发展和高级安全工程专业人才的培养起到十分积极的推进作用，同时，也为安全生产领域众多实际工作者提高专业理论水平提供学习资料。当然，这是第一套基于专业技术平台课程体系的教材，尽管我们的编审者和出版者夙兴夜寐，尽心竭力，但由于安全工程学科具有理论上的综合性与应用上的广泛性相交叉的特性，开办安全工程专业的高等院校所依托的行业类型又涉及军工、航空、化工、石油、矿业、土木、交通、能源、环境、经济等诸多领域，安全工程的应用也涉及人类生产、生活和生存的各个方面，因此本套系列教材依然会存在这样或那样的缺点和不足，难免挂一漏万，诚恳地希望得到有关专家和学者的关心与支持，希望选用本套系列教材的广大师生在使用过程中给我们多提意见和建议。谨祝本系列教材在编者、出版者、授课教师和学生的共同努力下，通过教学实践，获得进一步的完善和提高。

“嘤其鸣矣，求其友声”，高等院校安全工程专业正面临着前所未有的发展机遇，在此我们祝愿各个高校的安全工程专业越办越好，办出特色，为我国安全生产战线输送更多的优秀人才。让我们共同努力，为我国安全工程教育事业的发展做出贡献。

中国科学技术协会书记处书记[⊖]
中国职业安全健康协会副理事长
中国灾害防御协会副会长
亚洲安全工程学会主席
高等学校安全工程学科教学指导委员会副主任
安全科学与工程类专业教材编审委员会主任
北京理工大学教授、博士生导师
冯长根

⊖ 曾任中国科协副主席。

前 言

安全是人类生存和发展的基本要求，是生命与健康的基本保障。随着科技进步和社会的飞速发展，安全问题越来越受到社会的关注。为了减少意外事故，保障安全、健康的生产条件和作业环境，形成了安全科学与工程学科。

随着经济全球化的推进，安全工程专业英语是安全工程专业的人才所应具备的基本能力，也是了解与借鉴国外安全科学与工程领域先进技术的重要工具。希望通过本书能够提高安全工程专业人才的专业英语能力。本书的编写与出版得到了北京科技大学教材建设经费的资助。

本书是在借鉴国内外同类教材的基础上，为适应新的教学改革需要而编写的。全书内容广泛，共20个单元，系统地介绍了安全科学与工程学科的相关内容，使读者对该学科有更清晰的认识。同时为提高读者的专业英语水平，系统地介绍了科技英语的翻译与写作技巧。

本书由北京科技大学土木与资源工程学院安全科学与工程系黄志安、张英华担任主编，高玉坤、王辉任副主编；具体编写分工如下：黄志安、孙传武和王树祎编写1～5单元和附录，高玉坤和孙传武编写6～10单元，张英华、孙传武和张歌编写11～15单元，王辉、杨锐和燕立凯编写16～20单元。

本书由黄志安统稿，北京科技大学土木与资源工程学院的金龙哲教授和中国矿业大学（北京）的王凯教授担任主审，两位教授对本书进行了全面、细致的审阅，提出了许多宝贵的修改意见和建议。在此对两位教授的辛勤工作表示衷心感谢！

本书在编写过程中参考了国内外安全理论、矿山安全、工业安全相关的科技文献和书目，以及相关的科技英语翻译与写作的书目，在此对原作者表示诚挚的谢意。

由于编者的学识水平有限，书中难免有不当与错误之处，敬请广大读者及相关专家批评指正。

编 者

目　录

Unit One

Safety Theory

Despite an enormous amount of effort and resources applied to security in recent years, significant progress seems to be lacking. Similarly, changes in engineering are making traditional safety analysis techniques increasingly less effective. Most of these techniques were created over 50 years ago when systems were primarily composed of electromechanical components and were orders of magnitude less complex than today's software-intensive systems. New, more powerful safety analysis techniques, based on systems theory, are being developed and successfully used on a large variety of systems today, including aircraft, spacecraft, nuclear power plants, automobiles, medical devices, and so forth. Systems theory can, in the same way, provide a powerful foundation for security. An additional benefit is the potential for creating an integrated approach to both security and safety.

1. *Accident Causation Analysis and Taxonomy (ACAT) Model*

Since the concepts of man-machine-media model was first proposed by T. P. Wright, it has had a profound effect on accident analysis and prevention. Afterward, Management and Mission were introduced and the 5M model was established. In consideration of the complexity of system failure, more system factors have been incorporated into 5M model. For instance, Miller summarized seven system safety factors, which are man, machine, media, management, time, cost, and information. Irani et al. proposed a variation "5M" model to evaluate the impact of human, process and technology factors on information system failure. Kozuba suggested that though many efforts had been made to prevent undesirable flight-related events, human factor, technical factor and organizational factor were still the main causes. Of all these systematic safety factor models, the initial 5M model is the most widely used one and has been generally accepted in many areas, especially in ***aviation*** domains. It is a structured method which describes the subjects of safety analysis.

For a long time, man, machine, media, management, and mission have been recognized as the main elements contributing to accidents. However, it is too ***vague*** to include failures caused by supervision, decision making, regulations, or safety attitudes into management failure. Traditional management factor is a general subject which cannot provide more detailed types of failure. Based on accidents review, we identified six system safety factors, which are Man (M), Machine (M), Management (M), Environment (E), Information (I), and Resources (R). Among these system safety factors, machine refers to hardware in plant including all kinds of instruments, equipment, or vehicles. Man, which is also called human, refers to on-site personnel like operator, maintenance worker, office stuff, installer, or field supervisor. Their duties are to implement the decisions from managers. Management refers to supervision or decisions made by managers from plant units, companies, agencies, or government. Information includes procedures, programs, methods, standards, regulations, or laws. Resources include training, experts, raw materials, fund, energy, or products. Environment does not mean the physical environment but a social environment because the physical environment like weather is beyond controllability. It usually includes safety culture, attitude, or issues left over by history. Take BP Texas ***refinery*** accident as an example, inadequate preliminary hazard analysis and mechanical integrity program are categorized into information failure. To prevent this type of failure, attention should be paid to program formulation and evaluation.

Different considerations are defined for each factor of the system to detail its potential risks. However, due to lack of standards for the interpretation of these factors, different reference presents varied considerations. It leads to poor consistency in application. Hence, a structured theory is needed to guide the establishment of subgroups. Control theory can describe factors' functions and their communications with a closed loop. Each component in a control structure indicates a particular function that one factor should complete. A simplified diagram of a control structure is shown in Figure 1-1.

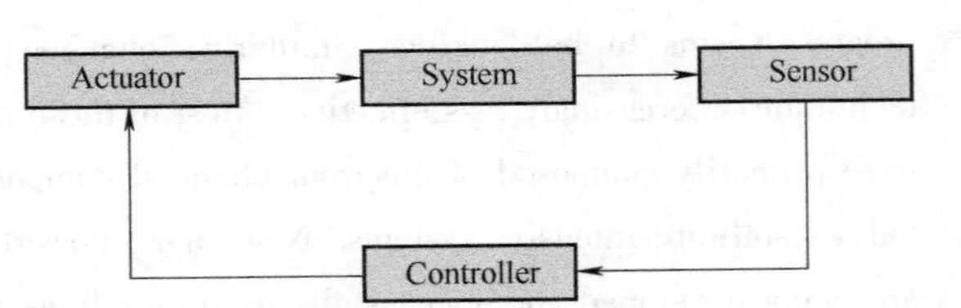

Figure 1-1 Simplified Diagram of a Control Structure

The basic components are actuator, sensor, and controller and communication, respectively. Therefore, from the perspective of control theory, it is assumed that each system safety factor has these four functional characteristics. The definitions of the functional characteristics are shown in Table 1-1. For example, if the mission is to open a valve, the control system can be described as follows: ① an operator who open the valve is actuator of the mission, ② sensor refers to the field supervisor whose job is to monitor the operation process, ③ an audit or evaluation should be made by a controller, and ④ all of these works require effective communications. Any missing function may ***lead to*** mission failure.

Table 1-1 Definitions of Control System Factors

Functional Factor	Function Definition
Actuator	Take measures or execute commands
Sensor	Measure and monitor the output
Controller	Compare output performance with the reference
→ Communication	Connect elements and convey information

To address the basic two issues of accident analysis, which are ① what is the failure and ② how does the failure happen, a new model is presented from both system safety perspective and control theory perspective. First, complex systems can be decomposed into six components, which are man, machine, management, information, resources, and environment from the view of system safety factors. From control theory perspective, actuator, sensor, controller, and communication are defined as system factors' functional abstractions. The combinations of system factors and control functions form a matrix model for accident ***causation*** analysis and classification, named Accident Causation Analysis and Taxonomy (ACAT) model. Then a comparison with existing cause classification schemes is made and the case of BP Texas refinery accident is used to illustrate its capability (see Table 1-2).

Table 1-2 ACAT Model and Elements' Definitions

Subject ➡ / ⬇ Function	Actuator (A)	Sensor (S)	Controller (C)	Communication (O)
Man (M)	H 11	H 12	H 13	H 14
Machine (M)	H 21	H 22	H 23	H 24
Management (M)	H 31	H 32	H 33	H 34
Information (I)	H 41	H 42	H 43	H 44
Resources (R)	H 51	H 52	H 53	H 54
Environment (E)	H 61	H62	H 63	H 64

No.	Description	No.	Description
H11	Fail to take effective actions	H41	Wrong or inadequate information
H12	Fail to monitor, or fail to detect the human failure in time	H42	Fail to monitor or update information
H13	Fail to follow procedures	H43	Fail to establish information
H14	Lack of effective communication between operators	H44	Fail to deliver or interpret information
H21	Design deficiency or malfunction	H51	Lack of training experts, raw materials, fund, energy or products
H22	Fail to monitor or detect the machine failure in time	H52	Fail to monitor the resource spending or changes
H23	Lack of sufficient machine maintenance	H53	Inadequate allocation of resources
H24	Information from equipment is not captured or interpreted	H54	Fail to deliver resources or resources needs
H31	Fail to manage workers or equipment or organization appropriately	H61	Ignore warnings or issues in previous events
H32	Fail to monitor organizational failure or manage change	H62	Fail to monitor the environment change
H33	Fail to follow procedure organizational inadequate decision	H63	No response to poor safety culture or attitude
H34	Lack of communication within decision levels	H64	Lack of communication culture

2. Reason's Swiss Cheese Model of Accident Causation

The systems approach is ***encapsulated*** in Reason's Swiss Cheese model of accident causation. The model states that in any system there are many levels of defense but these defenses are imperfect both because of inherent human ***fallibility*** and weaknesses in how systems are designed and operated.

Reason's model distinguishes between active failures and latent conditions. Active failures are errors and violations that are committed by people at the service delivery end of the system. Active failures by these people may have an immediate impact on safety.

Latent conditions result from poor decisions made by the higher management in an organization, e. g. by regulators, governments, designers, and manufacturers. Latent conditions lead to weaknesses in the organization's defenses, thus increasing the likelihood that when active failures occur they will combine with existing preconditions, breach the system's defenses, and result in an organizational accident. Latent conditions and active failures lead to windows of opportunity in a system's defenses. When these windows of opportunity are aligned across several levels of a system, an accident trajectory is created (see Figure 1-2). The accident trajectory is represented by the penetration of the levels of defense by an arrow. The holes represent latent and active failures that have breached successive levels of defense. When the arrow penetrates all the levels of defense, an adverse event occurs.

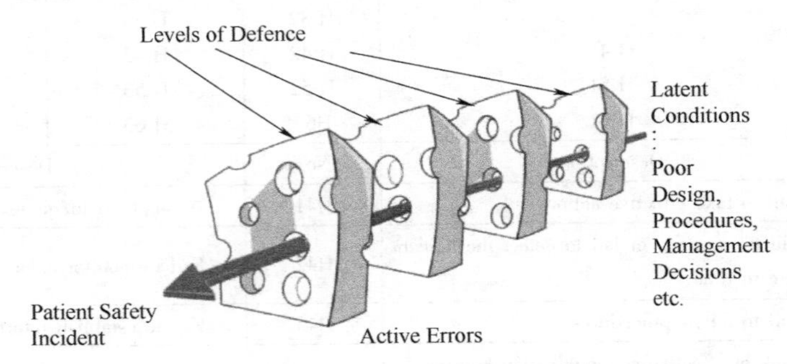

Figure 1-2 Reason's Swiss Cheese Model

3. Beyond Swiss Cheese

The Swiss Cheese model has been used as the theoretical basis for developing other models of incident causation and incident investigation tools in healthcare. The distinction between active and latent failures has strongly influenced efforts to understand the causes of error and incident investigation for the last two decades, both in healthcare and other industries. Its dominance has prevailed even though Reason himself has developed newer models ***aimed at*** understanding human error in complex systems. For example, the three buckets model and the harm absorbers model, both which recognize that healthcare professionals often use ***intuition***, expertise and foresight to anticipate, intervene and prevent patient harm.

4. The Old versus the New View of Human Error

Some critics have argued that, although well-intentioned, in practice, the Swiss Cheese model, leads to a linear approach to incident investigation: in what has been termed the old view of human error, efforts are made to trace back from active errors to identify organizational failures without recognizing the complexity of systems like healthcare and aviation.

Dekker distinguishes between the old view and the new view of human error. He argues that the old view of human error, where there is a search for organizational deficiencies or latent failures, simply causes us to relocate the blame for incidents upstream to senior managers and regulators. This was recently evidenced in the United Kingdom National Health Service in the Francis Inquiry reported into the deaths of patients at Mid Staffordshire hospital. There was a significant focus in both the inquiry report and in subsequent media coverage on the lapses by healthcare regulators that led to delays in intervening to prevent patients being harmed. As a result of the findings of the Francis Inquiry and other high profile national incident reports, the NHS's key regulator, the Care Quality Commission, has come under intense media scrutiny. Dekker's argument that blame is simply attributed further upstream seems, to some extent, to have been borne out by Mid Staffordshire.

Dekker advocates that a new view of human error is needed which views safety as an emergent property of a system in which there are numerous trade-offs between safety and other goals. Other theorists have also recognized that safety is an emergent property in complex systems, including proponents of ***resilience*** engineering.

5. Resilience Engineering

Resilience is the ability of individuals, teams and organizations to identify adapt and absorb variations and surprises on a moment by moment basis. Resilience engineering recognizes that complex systems are dynamic and it is the ability of individuals, teams and organizations to adapt to system changes that creates safety. Resilience moves the focus of learning about safety away from "What went wrong?" to "Why does it go right?".

One key concept from resilience engineering is the distinction between Safety I and Safety II. Safety has traditionally been defined by its absence. That is to say, we learn how to improve safety from investigating past events like incidents, complaints. This is known as Safety I. In contrast, Safety II focuses on the need to learn from what goes right. It involves exploring the ability to succeed when working conditions are dynamic. Safety II involves looking at good outcomes, including how healthcare organizations adapt to drifts and disturbances from a safe state and correct them before an incident occurs. We rarely learn from what goes right because resources are solely invested into learning from what goes wrong. However, serious incidents occur less frequently than instances of Safety II (which are numerous). Hence focusing on what goes right would provide an opportunity to about events that occur frequently, as opposed to rarely.

6. Amalberti's System Migration Model

Amalberti's system migration model is also relevant to understanding errors. Amalberti postulates that humans are naturally adaptable and explore their safety boundaries. A combination of life pressures, perceived vulnerability, belief systems and the trade-off between these factors versus perceived individual benefits leads people to ***navigate*** through the safety space.

Amalberti differentiates between: ① the legal space, i. e. prescribed behavior; ② the illegal-normal space, where people naturally drift into depending on situational factors and personal beliefs; ③ the illegal-illegal space; which brings people into an area of that is unsafe and where the probability of an accident occurring is greatly increased.

The legal-space is defined by policies, procedures and guidelines that describe standards of safe practice. Frequently, when serious incidents occur, non-compliance with policies and procedures is identified as a root cause. All too often hindsight bias comes into play in the investigation process and too little consideration is given to the situational factors that led to non-compliance. Hindsight bias occurs when an investigator, who is looking backwards after an incident has occurred, judges the behavior of those involved unfairly because with the benefit of hindsight it is easy to see the alternative courses of action that could have been taken which would prevent the incident from occurring.

7. Five Lessons about Safety and Accident Causation

Table 1-1 summarizes five lessons from the theories that have been summarized. It is ***postulated*** that future theories of safety need to take account of these five lessons in order to develop models and frameworks that capture the complexity of safety in healthcare. Without an under-pinning theoretical framework that captures how safety is a complex, dynamic phenomenon, healthcare organizations around the world will not understand the different facets of safety that emerge as healthcare systems evolve over time (see Table 1-3).

Table 1-3 Five Key Lessons from Previous Theories of Accident Causation, Human Error, Foresight, Resilience and System Migration

What is the lesson to learn?	Source 1
Lesson 1: A combination of systems and human factors can enhance or erode safely.	Swiss Cheese; Three buckets and harm absorber models
Lesson 2: Systems are dynamic: they evolve over time and spring nasty surprises. Healthcare professionals, teams and organizations sometimes successfully anticipate and manage these nasty surprises, and sometimes they do not.	Three buckets and harm absorber models; Resilience engineering; Safety Ⅰ *versus* Safety Ⅱ
Lesson 3: Safety is an emergent property of the system which needs to be understood in the context of trade-offs with other competing goals (for example, in healthcare, meeting efficiency targets, making financial savings and ensuring continuity of the service).	The old and new view of human error; Resilience engineering
Lesson 4: Hindsight bias, together with the human tendency to attribute blame and the fact that serious incidents occur less frequently than successful outcomes limits what we can learn from taking human error as our starting point and tracing backwards to identify the causes of what went wrong. We therefore need to balance our focus and learn from what goes right rather than being preoccupied with learning from what goes wrong.	The old and new view of human error; Safety Ⅰ *versus* Safety Ⅱ
Lesson 5: Humans migrate and explore the system's safety boundaries. The extent to which they do this depends upon a combination of factors including life pressures, situational factors and personal belief systems.	System migration

8. The Safety Evolution Erosion and Enhancement Model

The five lessons summarized in Table 1-1 are illustrated in Figure 1-2. It shows the underlying processes that healthcare organizations need to appreciate to understand safety as a complex, dynamic process. It shows how system evolution impacts on safety both positively and negatively by causing both erosion and enhancement. Figure 1-2 shows that both systems-level, team and individual human factors can enhance or erode safety in healthcare. Figure 1-2 aims to show that any system, whether it is a healthcare system, aviation, offshore oil and gas, banking or other complex sociotechnical system naturally evolves over time. System evolution is caused by many different types of factors including the introduction of new technology, innovations in healthcare procedures, organizational restructuring or mergers, staff retention and recruitment, patient pathway and process redesign, the context and focus of external regulators, cultural change, equipment maintenance and IT upgrades, production, efficiency and safety performance targets, and the economic climate. System evolution also occurs as a result of the natural human tendency to explore safety boundaries (see Figure 1-3).

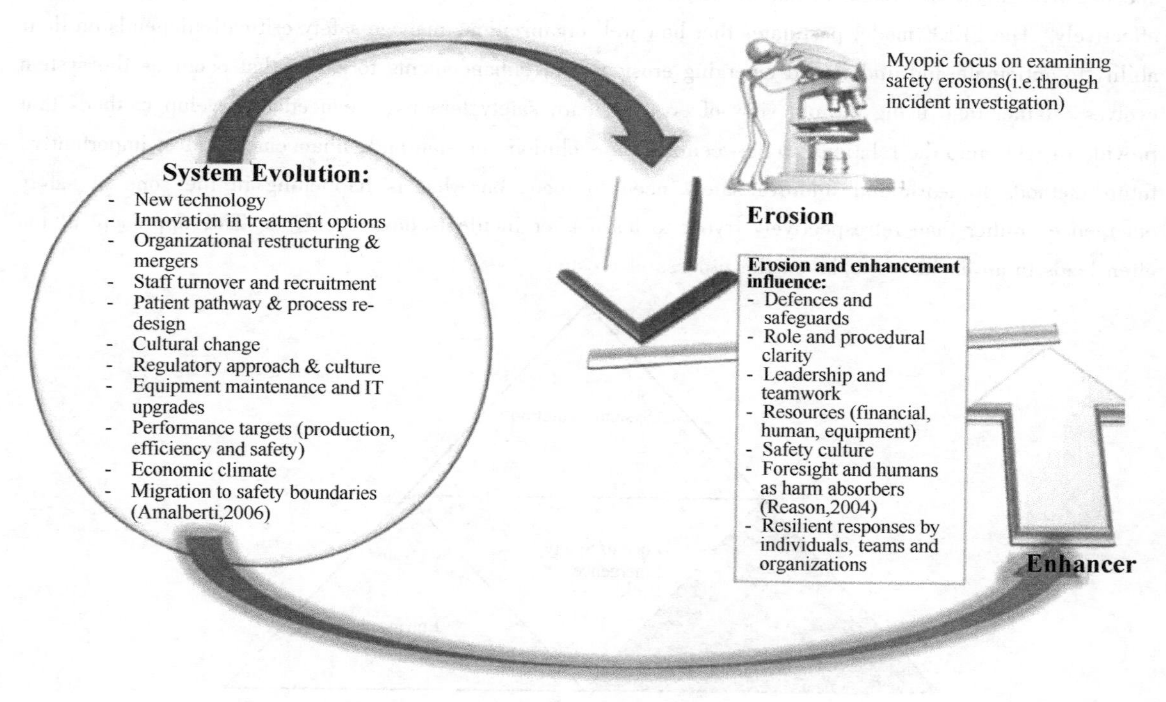

Figure 1-3 The Processes of Safety Evolution, Erosion and Enhancement

(1) Enhancement and Erosion

System evolution can have both positive and negative effects on safety. ***In terms of*** positive effects, system evolution can lead to safety enhancement. Enhancement occurs when system evolution strengthens in-built defenses, barriers and safeguards or where it improves the ability of individuals and teams to anticipate and respond when nasty surprises occur. By recognizing that safety enhancement results from system evolution, Figure 1-2 illustrates the importance for organizations of learning from what goes right, rather than only learning from what goes wrong.

System evolution can also lead to erosion, where defenses, barriers and safeguards are weakened or where the ability of teams and individuals to identify, intervene and thwart emerging safety threats is negatively affected (see Figure 1-2).

The central tenet of the SEEE model is that in order to effectively manage safety, healthcare organizations need to understand the relationship between system evolution and enhancement and ***erosion***. Hence we need to recognize that safety is an emergent property of the system and that it needs to be understood in context of how it is balanced against other competing goals like production and efficiency targets.

Organizations who only focus on learning about system erosion invest time and resources learning from past harm and some elements of reliability without paying sufficient attention to the other dimensions of the measurement and monitoring framework.

(2) Understanding Safety Emergence

The SEEE model is shown in Figure 1-4. Figure 1-4 shows how system evolution leads to both erosion and enhancement. Importantly, it also shows the zone of safety emergence. This is the area where positive and negative effects of system evolutions need to be anticipated and understood in order to manage safety effectively. The SEEE model postulates that how well organizations manage safety critically depends on their ability to anticipate and understand emerging erosions and enhancements to safety that occur as the system evolves. Rather than using the old view of error to learn safety lessons, we need to develop methods that provide insights into the relationship between system evolution, erosion and enhancement. Most importantly, future methods to learn and improve safety need to focus on what is happening in the zone of safety emergence, rather than retrospectively trying to learn after incidents have occurred. This approach all too often leads to an exclusive focus on the process of erosion.

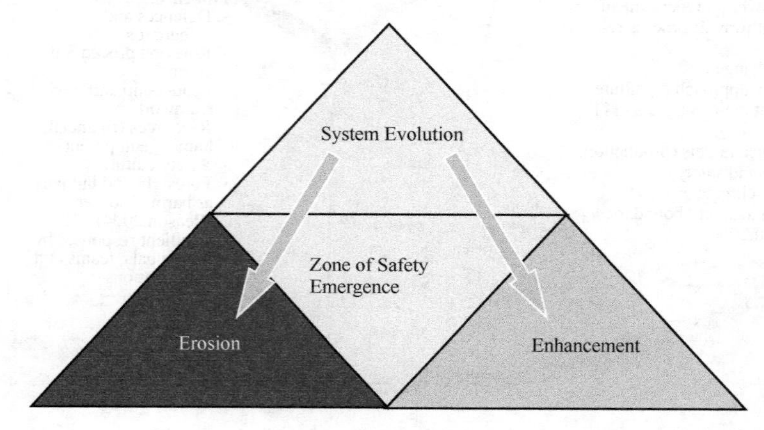

Figure 1-4 The System Evolution, Erosion and Enhancement Model of Safety

New Words and Expressions

aviation [eɪvɪˈeɪʃ (ə) n] n. 航空，航空制造业，飞行

vague [veɪg] adj. 含糊的，不明确的，不清楚的，含糊其辞的

refinery [rɪˈfaɪn (ə) ri] n. （石油等的）精炼厂

lead to 导致；通向

causation [kɔːˈzeɪʃ (ə) n] n. 原因，诱因，起因

encapsulate [ɪnˈkæpsəleɪt] v. 压缩，装入胶囊

fallibility [ˌfæləˈbɪləti] n. 易错，不可靠，易误

aim at 针对；目的在于

intuition [ɪntjʊˈɪʃ (ə) n]	*n.* 直觉力，(一种) 直觉
resilience [rɪˈzɪlɪəns]	*n.* 适应力，弹力，快速恢复的能力，还原能力
navigate [ˈnævɪgeɪt]	*v.* 导航，航行，航海，横渡
postulate [ˈpɒstjʊleɪt]	*v.* 假定，假设
in terms of	依据；按照；在……方面；以……措辞
erosion [ɪˈrəʊʒ (ə) n]	*n.* 腐蚀，糜烂

Exercises

1. What is safety theory? What is systems theory? What is safety analysis techniques?
2. Explain the relationship between safety theory and systems theory.
3. What does the safety analysis method contain?
4. Explain the positive and negative effects of system evolution on safety.
5. Indicate the functions of safety analysis techniques by using practical cases.

科技英语的阅读方法和技巧

科技英语相比于普通英文文章来说，在语言、词汇、题材内容上都有难度，需要对文中的信息进行综合加工、概括、归纳，然后得出结论。

1. 阅读方法

阅读时，要注意以一个意群为单位，将一个较长较难的句子划分为一个个小单元，方便理解。阅读方法有略读、浏览、精读。

（1）略读

略读是指快速地浏览文章，了解文章的主旨大意，对文章内容有总体的印象。段落的开头和结尾往往代表段落主旨，文章的首段和尾端往往说明文章的主题思想，因此，在略读时，这些重点位置必须看，从而掌握主旨信息，细节部分可以不读。

（2）浏览

浏览就是在文中找寻某一问题、某一观点或关键的词语。浏览时，快速地扫视全文，找到相关信息，尤其是一些有特点的关键词语，从而确定目标范围，不相关的内容可以略过。浏览可以帮你只阅读特定的信息而省去阅读无关内容的时间。

（3）精读

精读是指仔细阅读文章，包括作者的意图、中心思想的掌握，逐字逐句对文章进行深入的理解。

2. 阅读技巧

科技英语文献在不同的领域有不同的阅读技巧，但对于科技文献本身是存在一定规律的。

（1）紧抓主题思想

每一篇文章都有一个主题思想，要准确地掌握主题思想，就要找到文章的主题名词、主题句。在获得主题思想时，不能以偏概全，要通读全文，归纳总结，找出能概括全文思想的关键词与关键句。

（2）获取文章细节

在阅读文章时，为了更好地理解文章内容，要注意能够表达作者观点的细节部位。细节往往就是一个例子或者事实，要准确区分细节和主旨的区别。

（3）推敲生词含义

科技英语文献中，会出现许多生词，这时通过查字典解决不了问题，那么就需要通过对上下文的理解来推断生词的含义。例如，出现“or”“and”等词语时，前后词语意思相近，出现“but”等具有转折性的词语时，前后词义相反。

（4）了解文中指代关系

科技英语中经常使用it等指代性的名词，下面举例说明it的用法：

1）it用作代词，指代无生命的东西、物体及抽象概念，也可指代前面出现过的名词。

This project is dangerous. It is not safety.

这个工程危险。它不安全。

2）it用在习语中的组成部分及与天气、时间有关的非人称主语。

It is sunny today.

今天是晴天。

It is raining all the weekend so that we cannot have an outing.

周末一直在下雨，以至于我们不能外出。

3）it用作形式主语，起先行作用，没有具体意义。

It is certain that most engineering materials are partly elastic and partly plastic.

可以肯定，大多数工程材料有几分塑性，也有几分弹性。

It is obvious that these techniques will be of great value in cryogenic switching circuit.

很显然，这些工艺对低温开路有很大的价值。

（5）把握文章的对比关系

作者通常会将读者比较熟悉的概念或事物与所说内容做对比，方便读者理解所述内容。

Reading Material

Safety Climate: New Developments in Conceptualization, Theory, and Research

In 2007 there were around 4 million nonfatal workplace injuries and illnesses in the U. S., which according to the U. S. Bureau of Labor Statistics (BLS), occurred at a rate of 4. 2 cases per 100 equivalent full-time workers. That same year there were also 5657 fatal occupational injuries (United States Department of Labor, 2009). These statistics are concerning, especially since the BLS has been shown to undercount injuries associated with chronic or acute conditions. These data illustrate the continuing need to identify ways to reduce workplace accidents and injuries and to improve overall workplace safety. Efforts to increase workplace safety encompass several areas. Standard interventions employed to reduce risk can be grouped within four broad categories: engineering (e. g., redesigning a tool or installing machine guards); ***administrative*** (e. g., changing job procedures or rotating workers through a particular job); personal protective equipment (e. g. , protective glasses or hearing protection); and education and training. Although there is general agreement that both management leadership and employee participation are critical to the success of reducing injury risk, until recently, most intervention efforts have relied on a traditional engineering approach.

In an attempt to make further improvements, risk managers and safety directors are exploring organizational and psychosocial factors in the workplace to complement other approaches. One of the most prominent factors currently under consideration is that of safety climate.

Safety climate is an organizational factor commonly cited as an important ***antecedent*** of safety in the workplace. Zohar was one of the first to introduce the concept of safety climate as a way to describe employees' perceptions of the value and role of safety within their organizations. Specifically, safety climate ***refers to*** the workers' perceptions of the organization's policies, procedures, and practices as they relate to the value, importance, and actual priority of safety within the organization. The practical and theoretical significance of safety climate as a construct derives from its ability to predict safety behavior and safety-related outcomes (e. g. , workplace accidents and injuries) in a wide variety of settings, and across cultures.

Theoretically, safety climate provides a framework to guide the safety behavior of employees in such a way that they develop perceptions and expectations regarding safety behavior outcomes and, thus, behave accordingly. In large part, workers develop these perceptions and expectations about the priority and importance of safety by observing the actions of their supervisors. Safety climate predicts employees' motivation to work safely, which affects employees' safety behaviors and subsequent experiences of workplace injuries or incidents. It has also been directly associated with increases in safety behaviors and decreases in workplace injuries.

1. Introduction

In the "high reliability" industries, where significant hazards are present (even if rarely realized), operating companies and their regulators pay considerable attention to safety assessment. In recent years there has been a movement away from safety measures purely based on retrospective data or "lagging indicators" such as fatalities, lost time accident rates and incidents, towards so called "leading indicators" such as safety audits or measurements of safety climate. It can be argued that these are predictive measures enabling safety condition monitoring, which may reduce the need to wait for the system to fail in order to identify weaknesses and to take ***remedial*** actions. This can also be conceptualized as a switch from "feedback" to "feed forward" control.

The shift of focus has been driven by the awareness that organizational, managerial and human factors rather than purely technical failures are prime causes of accidents in high reliability industries. The nuclear power industry recognized the importance of organizational culture following the Chernobyl accident and encouraged operators to assess the safety culture on their plants. The idea of a safety culture is predated by an extensive body of research into organizational culture and climate, where culture embodies values, beliefs and underlying assumptions, and climate is a descriptive measure reflecting the workforce's perceptions of the organizational atmosphere. Longstanding debates as to the nature, ***supremacy*** and applicability of culture versus climate in organizational theory are now being echoed by the safety researchers. Cox and Flin reviewed some of the arguments and concluded that in terms of operationalizing the concept into a practical measurement tool for managers, safety climate was the preferred term when psychometric questionnaire studies were employed as the measurement instrument. Safety climate can be regarded as the surface features of the safety culture ***discerned*** from the workforce's attitudes and perceptions at a given point in time. It is a snapshot of the state of safety providing an indicator of the underlying safety culture of a work group, plant or organization. If this concept is to be effectively translated into an operational measure for safety management then a number of questions need to be addressed. What are the key features of a good safety culture that can be assessed by a climate measure? Can these be regarded as generic features of the safety culture or are they specific to certain companies, industries or cultures? Is there any evidence that these features are indicative of the state of safety, for instance do they relate to other safety measures (e. g. accident rates)?

The International Atomic Energy Authority provides a set of safety culture indicators in the form of a question set. Those that relate to operations (rather than design or regulation), cover definition of responsibilities, training, management selection, reviews of safety performance, highlighting safety, workload, relation between management and regulator, management attitudes, individual attitudes, local work practices and supervision. A British advisory committee on human factors in nuclear safety identified senior management commitment, management style, management visibility, communication, pressure for production, training, housekeeping, job satisfaction and workforce composition as key indicators of the safety culture. It recommended the assessment of safety climate using a survey approach, advice now endorsed by the UK safety regulator. For managers and researchers selecting a safety climate measure, is there a common set of organizational, management and human factors that are being regularly included in measures of safety climate?

Recent academic interest in the measurement of safety climate, has resulted in a ***proliferation*** of assessment instruments, typically in the form of self-report questionnaires administered as large-scale surveys in different sectors, principally the energy industries, but also in manufacturing and construction. It could be argued that the lack of a unifying theoretical model in this area is a reflection of the state of development of this field, where an inductive rather than a ***deductive*** approach is in operation. However, it does mean that these instruments tend to have distinct developmental histories, often customized to the sponsoring organization's requirements. In the main, they are designed to measure a set of themes derived from reviews of the safety research literature.

Interviews and focus groups conducted at the worksite are used to reveal particular issues concerning the workforce and to tailor the instrument accordingly. Only a few independent replications of questionnaires and examination of the resulting factor structures have been undertaken. The recent emergence of a number of new scales seems to have triggered efforts to address this problem by comparing safety climate scales from different studies. The initial reviews demonstrated that measures vary significantly in almost all respects—content, style, statistical analysis, sample size, sample composition (workers, supervisors, managers), industry and country of origin. Factor analysis is typically used for identification of an underlying structure but numbers of items range from 11 to 300 and thus solutions range from 2 to 19 factors. Drawing direct comparisons between factor labels and (loading) items across these measures remains problematical due not only to the methodological inconsistencies outlined above, but also to cultural and language differences across both countries and industries. Williamson et al. examined seven reports measuring safety climate and concluded that eight factors could be discerned, four measuring attitudes and four perceptions, although they presented no detailed analysis of which of these eight factors were derived from particular questionnaires. Dedobbeleer and Beland reviewed safety climate instruments and argued that only two factors, management commitment and worker involvement, had been properly replicated across studies. Coyle et al. found different factor structures using the same safety climate scale in two Australian health care organizations and concluded that the likelihood of establishing a universal and stable set of safety climate factors was "highly doubtful". Thus we have very limited evidence for or against a common set of core features, notwithstanding a prevailing belief that, "a specific example of good practice may not always be directly transferable, unlike the underlying features and attributes which are universally applicable". As the number of scales multiplies, a superficial scrutiny of their component themes does suggest that a basic set of features is beginning to emerge. Thus the field may be moving towards the position where a base ***taxonomy*** of fundamental safety climate attributes could be distilled from the proliferation of scales and items, ***akin to*** the "Big Five" factors in personality measurement. To test this proposition, a larger sample of safety climate studies was examined with particular reference to the composition of each questionnaire and its validation.

2. *Summary and Prospect*

The number of studies on safety climate has increased dramatically in recent years from the first one in 1980 to a total of 130 articles published in peer-reviewed journals through 2008 (similar to the figure in Glendon on safety culture/climate, distribution of these 130 safety climate articles by year is illustrated in Figure 1-5). Researchers from various countries have begun paying attention to it.

Although the important role of safety climate in safety outcomes has been established and many studies have been done by scholars in different disciplines, there are still gaps in the literature. Three specific areas of safety climate still should be paid attention to in order to help fill the gaps and give a focus to the research.

The first theme of the special issue is on the topic of new developments in the conceptualization of safety climate. Possible themes or questions to explore: what are safety climate and safety culture and how do they differ; what should safety climate dimensions encompass; how is safety climate formed or developed; how should safety climate scores be validated.

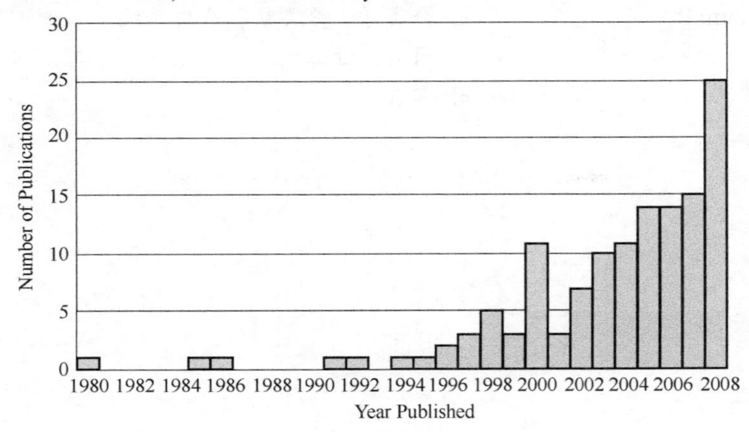

Figure 1-5 Year of Publication of 130 Peer-Reviewed Journal Articles on Safety Climate

(*Note*: APA Psyc NET search engine, which included both psychology and safety journals, was used to search for the topic "Safety Climate" in abstracts of peer-reviewed journal articles.)

Further, although extensive prior research has shown that safety climate predicts safety-related outcomes, relatively few studies have conducted interventions that attempt to improve safety climate. Therefore, the second topic is to provide a sampling of the most recent research being conducted on interventions to improve safety climate.

Finally, in the safety climate literature, some studies have been conducted within a single organization or industry, whereas other studies have focused on workers across work settings. It is important to contrast safety climate research conducted within versus across occupations or industries. Thus, the final topic of the special issue is on occupation/industry-focused studies of safety climate, with the goal of identifying different challenges and findings that arise within or across various occupations or industries.

A multilevel study examined the role of safety climate on accident under-reporting, bringing the theory of planned behavior and a new analytical approach (dominance analysis) into the equation.

Several articles addressed safety climate in different industries, such as the rail industry, the ***retail*** trade, the industrial workplace, health care, the petroleum sector, and the manufacturing sector. However, no appropriate studies were identified in terms of interventions to improve safety climate. Hence, more studies on interventions which target on improving safety climate for the industries should be encouraged. Also important are new studies which target on the predictors of safety climate which can serve as potential factors for interventions.

New Words and Expressions

administrative [əd'mɪnɪstrətɪv] *adj.* 管理的，行政的
antecedent [ˌæntɪ'siːd (ə) nt] *n.* 先行词，前事，前情，祖先
refer to 引用
remedial [rɪ'miːdɪəl] *adj.* 旨在解决问题的，补救的，纠正的，辅导的
supremacy [s (j) uː'preməsi] *n.* 至高无上，最大权力，最高权威，最高地位
discern [dɪ'sɜːn] *v.* 看见，辨明，觉察出，识别
proliferation [prəʊ-ˌlɪfə'reɪʃən] *n.* 增生，〈正式〉(数量) 激增，扩散
deductive [dɪ'dʌktɪv] *adj.* 演绎的，推论的，推理的
taxonomy [tæk'sɒnəmi] *n.* 分类学，分类法，分类系统
akin to 同类，近似
retail ['riːteɪl] *n.* 零售

Unit Two

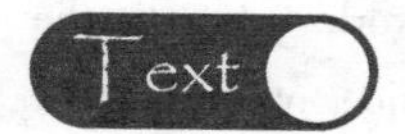

Safety Economics

Occupational injury and illness are matters of health, but they are also matters of economics, since they stem from work, and work is an economic activity. The economic perspective on Occupational Safety and Health (OSH) ***encompasses*** both causes and consequences: The role of economic factors in the etiology of workplace ill-health and this has the effects on the economic prospects for workers, and enterprises.

Broadly speaking, there are three general purposes that economics can serve for OSH. First, identifying and measuring the economic costs of occupational injury and disease can motivate the public to take these problems more seriously. Second, understanding the connections between the way firms and markets function and types of OSH problems that arise is ***crucial*** for the success of public policy. Why are conditions better in some sectors or regions than others, and why are particular groups of workers at greater risk? As the pace of economic change picks up throughout the world, these questions need to be addressed on a continuing basis. Finally, as important as the protection of worker health and well-being is, it is not the only objective of modern society. Economic analysis can help show when safeguarding working conditions are ***complementary*** to other social goals, and it can ***illuminate*** the tradeoffs when it is not.

1. Cost of Safety

For all of these goals, a central concept is that of costs. On the one side, we have the costs of improving the conditions of work, in order to reduce the incidence of injury and disease. On the other side, we have the costs of not doing these things. But the concept of costs is not simple; there are many kinds of costs.

The total costs of accidents come under two major headings:

1) Those covered by insurance.

2) Those not covered by insurance.

The costs covered by insurance are those which come within the scope of workers' compensation, motor vehicle, property, machinery damage, fire, public risk, etc. All premium rates are based on the basic principles of insurance:

- the risk and extent of potential loss;
- past experience.

The costs of accidents not covered by insurance are high, and can be several times the cost of those covered by insurance. Following on-the-job accidents, there will be disruption of the orderly processes, and these interruptions must be paid for. Some of the costs involved.

(1) The Costs of Wages for Time Lost by People Not Injured

Ordinarily, employees near the scene of an accident which results in injury or damage will stop to watch, assist or discuss the accident. They may not be able to continue without the aid or output of the injured worker or damaged equipment.

It is reasonable to assume each employee will be producing a given amount in work value to the organization every hour. This will be at least as much as the wages the employee is paid; otherwise the employee would not be there. In other words, the person has to be producing (in the U. S. currency, for example) $ 20.00 before being paid $ 20.00. Obviously this is the absolute minimum. In addition, ***allowance*** has to be made for overheads, power, space, administration and the like. Close examination will show that this amount will often be two to ***two-and-a-half times*** the amount of wages paid before a start can be made on profits; that is, $ 40.00 – $ 50.00 per hour.

(2) The Costs to Repair or Replace Property Damage

These relate to the costs of repairing or replacing machinery, raw materials, work in process and property not insured, inadequately insured or not insured for replacement value. There are difficulties in estimating the real worth of an old machine destroyed by accident. There are many machines in use today in which all, or practically all, of the original price has been ***written-off*** in depreciation. This does not mean that the destruction of such machines is of no loss to the organization. If a machine was bought ten years ago for $ 10,000 and at the date of destruction was still in good operating condition, even though the book value was nil, the purchase of a replacement machine to bring the plant to the same operating efficiency could now cost $ 16,000. The loss would be $ 16,000 and any additional ***taxation*** advantages such as depreciation, less the recoup of any salvage value and insurance payouts.

(3) Loss of Production by the Injured Worker

As in the case of the persons not injured, the work value of the injured person during time lost must be regarded as worth at least as much to that person's employer as the amount of wages paid for the employee's time. The loss also occurs when the employee is away from the job being treated for minor injuries, with the added cost of actual wages paid for the period.

(4) Cost of Supervisor's Time

Supervisor's time is taken up following accidents, making adjustments, arranging assistance for the injured workers, preparing accident reports, training replacement workers and rearranging staff and workloads. The organization loses the value of the work the supervisor would have performed during the period of "cleaning up" after an accident. Supervisors do not just sit around waiting for an accident to occur to make themselves useful. In addition to getting work out on time, they are responsible for planning, instructing workers, improving methods, eliminating bottlenecks and the like. If supervisors had more than adequate time then the ratio of supervisors to workers could be reduced.

(5) Decreased Output of the Returned Injured Worker

Often workers are not fully recovered when they return to work after being injured, and do not work at their normal rate. The percentage of pay which corresponds with the percentage reduction in output is a loss.

(6) Uninsured Medical Costs Borne by the Organization

This cost is for the services supplied and equipment used in maintaining a first aid unit. Bandages, medication, vehicles and buildings have to be paid for, as well as the cost of maintaining the first aid centre and attendants.

(7) The Cost of Time Spent by Senior Staff

Accidents have to be investigated, particularly those involving serious injury or damage. As a result activities such as meetings, reviews, follow-ups and other discussions occur. These often involve senior staff. This represents a loss of productive work value.

(8) Other Costs

The costs mentioned above represent only a few of the valid elements of accident costs. Other costs include such items as hiring replacement staff and/or equipment, excess spoilage of product, damaged minor equipment, the cost of preparing reports for government departments and insurance companies, and many more. In the case of, for example, construction companies tendering for projects, serious accidents can result in loss of future contracts.

(9) Cost of Improving Level of OSH

Firms will endure the burden of engaging in costly OSH prevention and risk assessment efforts provided that the expected benefits outweigh the immediate outlays. Those measures of prevention include the following aspects: enhancing workplace ergonomics, purchasing of personal protective equipment, OSH training to employees, safety management and inspections, and so on.

2. Theoretical Underpinnings for Determining the Level of Safety

The level of OSH in an economy is determined by the interplay of incentives faced by workers and firms in job and insurance markets and is ***moderated*** by the regulatory activities of government. Economists have typically analyzed the safety decisions of both employees and firms in a perfectly competitive market as the outcome of a complex set of factors reflecting various costs and benefits of investing in OSH. On the one hand, workers are assumed to be rational and fully informed agents who are likely to demand a "wage premium" as compensation for OSH risks. Their safety activities on the job and their degree of absenteeism once ill or injured is also believed to be affected by their exposure to social insurance and compensation benefit schemes. On the other hand, employers are assumed to face a tradeoff between the cost of investing in health and safety precautions and the expected benefits of such actions, most notably lower wages offered to employees as compensation for job disamenities or risk. The equilibrium level of OSH in the economy is therefore the outcome of the interaction between the labour demand and labour supply decisions of firms and workers, respectively.

(1) Heterogeneity in Market Opportunities Offered by Employers

A convenient formalization of the safety decisions faced by employers in market economies is shown in Figure 2-1. In the absence of regulation, a sufficient condition for profit-maximizing firms selecting the optimal safety level, S_f^* (or, equivalently, the optimal job risk level, R_f^*), is that their OSH decisions should equalize their OSH costs and OSH benefits at the margin. The ante-factum outlays associated with improvement in workplace conditions (e.g. preventative practices, OSH training) are likely to rise as the level of health and safety increases. This is reflected in the convexity of the total costs curve in Figure 2-1. The post-factum benefits (or, equivalently, reduced damage costs) are associated with having to pay lower (compensating) wages and workers' compensation due to the reduction of workplace injuries and illnesses, a lower level of sickness absence and sick pay, reduced insurance premiums, and the evasion of other non-trivial costs to the firm (disruption in production, replacement of specifically trained workers, low worker morale, bad reputation, etc.). These ***are likely to*** fall at a diminishing rate as OSH levels increase, as shown by the downward-sloping total benefits curve, because the severity of incidents at higher levels of safety are subdued. As illustrated in Figure 2-1, only at S_f^* (R_f^*) is the cost of extra safety equalized with the value of the benefits that an inframarginal unit of job safety entails for a given firm.

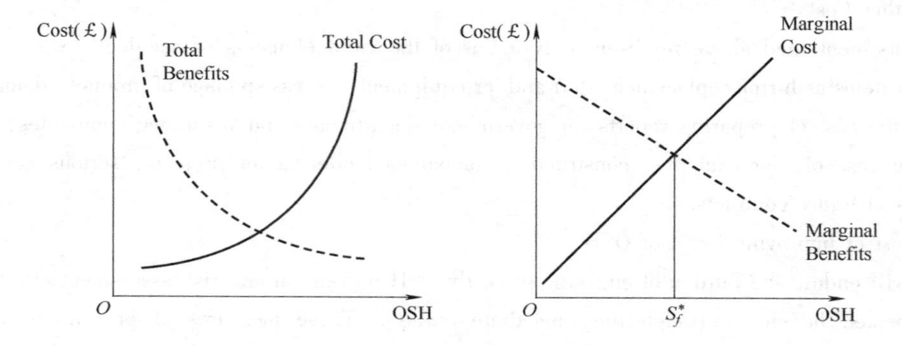

Figure 2-1 Efficiency in the Market for OSH

The above reasoning implies that in order to maintain constant profits, firms must offer lower wages (w) to offset any marginal addition in the cost of OSH. The wage offer of a given firm is therefore an increasing function of job risk, as is depicted by the concave is profit curve *II* in Figure 2-2.

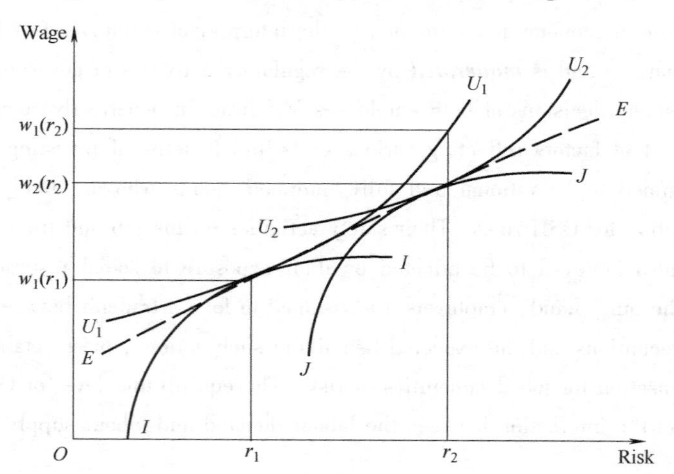

Figure 2-2 Equilibrium in the Market for OSH

Furthermore, the optimal wage-risk pairs (R_f, W_f) that can maintain the same level of profits are likely to vary across different firms, given that they face a diverse set of technologies and job hazards. Working environments differ with respect to the exposure of employees to various inputs of the production process. So it is clear that firms differ in their ability to tradeoff the costs of providing a safer working environment with the expected benefits (particularly, a lower wage bill). This is illustrated by the heterogeneity in the wage offer curves *II* and *JJ* of two hypothetical firms in Figure 2-2.

(2) Heterogeneity in Preferences of Workers for Job Risk

On the supply side of the market it is assumed that rational workers, who have perfect information and exhibit a high degree of labour market mobility, are likely to demand a wage premium as compensation for their willingness to endure danger on the job that is not covered by their ex-post accident or sickness insurance. In order to maintain expected utility constant, a given worker will therefore demand alternative levels of wage compensation for varying degrees of job risk, as summarized by the U_1U_1 curve in Figure 2-2. However, workers exhibit differential tolerances to risk for a variety of reasons related to economic and demographic circumstances or simply differences in tastes. This heterogeneity in workers' willingness to be employed in disagreeable job environments can therefore be depicted in Figure 2-2 by the dissimilar expected utility loci of two hypothetical workers, U_1U_1 and U_2U_2.

(3) Compensating Wage Differentials (CWDs) for Job Risk and Other

Disamenities Superimposing the labour supply choices of individuals on the wage offer curves of firms results in what is known as "the fundamental long-run market equilibrium construct" of labour economics, namely the theory of CWDs. Rooted in the insightful work of Adam Smith's *Wealth of Nations*, the theory predicts that market forces will ensure the payment of wage premiums by firms which are characterized by ***inferior*** working conditions, as a means of recruiting and retaining valuable labor. It is further postulated that in a perfectly competitive labour market a positive equilibrium wage-risk relationship (shown by line *EE*) should arise due to the "matching" of the preferences of workers and firms. This is shown by the points of tangency between the expected utility loci and the market wage opportunity curves, (r_1, w_1) and (r_2, w_2), in Figure 2-2. Risk-averse workers are thus expected to take up jobs in firms which find it optimal to provide a safer work environment, whereas less risk-averse workers are more willing to be employed by firms which face a dearer marginal cost of safety provision. Such an assortative matching procedure predicts that jobs characterized by a higher degree of job risk (or other disamenities) should, in equilibrium, offer compensating wage rents, ceteris.

New Words and Expressions

encompass [ɪn'kʌmpəs]	v. 围绕，包围，包含，包括
crucial ['kruːʃ(ə)l]	adj. 至关重要的，关键性的
complementary [kɒmplɪ'ment(ə)ri]	adj. 互补的，补充的，相互补足的
illuminate [ɪ'l(j)uːmɪneɪt]	v. 照亮，照明，阐明，解释
allowance [ə'laʊəns]	n. 津贴，补贴，免税额，限额
two-and-a-half times	两倍半
written-off	已注销
taxation [tæk'seɪʃ(ə)n]	n. 税收，征税，税制，课税
moderate ['mɒdəreɪt]	v. (使)减轻，缓和，节制
be likely to	有可能，很可能
inferior [ɪn'fɪəriə]	adj. 较差的，次的，比不上……的，级别低的

Exercises

1. What is occupational safety and health?
2. What are the three general purposes that economics can serve for OSH?
3. Describe the theoretical underpinnings for determining the level of safety.
4. What does the safety cost contain?
5. What are the difficulties in estimating the real worth of an old machine destroyed by accident?

科技英语的语言特点

科技英语是为表达科技概念、理论和事实的英语，主要呈现方式为教材、论文、综述、专利、科技新闻、实验报告、说明书等，在词汇、句法结构和语篇等方面都有自己的特点，下面分别介绍。

1. 词汇

科技英语在用词方面最主要的特点表现在使用大量的专业词汇，构词方式灵活多变，以及对日常词汇的借用等方面。

（1）大量使用专业词汇

科技英语因其特殊性，主要表达的是科技概念、理论和事实，专业性强。例如：

The *air pressure* at the *hot valve inlet* does not become excessively low.

在热风阀进口处的风压不要变得太低。

The *water vapor* will change from its invisible state to *condense into* visible *moisture* when the *dew-point* temperature is reached.

当露点温度达到时，水气将从其看不见的状态凝结成可见的水分。

（2）灵活的构词方式

随着科技的不断发展，在科技英语的用词方面也有了创新，有了灵活多变的构词方式。例如：

1）合成词。

splashdown 溅落

salt-former 卤素

fallout 放射性尘埃

dew-point 露点

hot-press 热压机

pulse-scaler 脉冲定标器

2）混成词。

contrail = condensation + trail 凝结尾流

gravisphere = gravity + sphere 引力作用范围

teleprinter = teleprinter + exchange 电传

3）词缀词。

anti-smog 反烟雾

anti-particle 反粒子

anthropology 人类学

4） 缩略词。

FTA（Fault Tree Analysis）事故树分析

FMEA（fault Mode and Effects Analysis）失效模式与效应分析

PHA（Preliminary Hazard Analysis）预先危险性分析

ETA（Event Number Analysis）事件树分析

HAZOP （Hazard and Operability Analysis）危险与可操作性研究

5）外来词。

Kuru 库鲁瘢痫症

Ohm 欧姆

Alfven wave 阿尔文波

（3） 借用日常词汇

illustrate 图解，举例说明，阐明

underlying 根本的，潜在的

methodology 方法学，方法论

abatement 消除，减少，减轻

2. 句法结构

科技文献的句法结构中多使用名词化结构、被动句、非限定性动词、后置定语、长句，较少使用人称代词和描述性词汇，从而造成句式的复杂。

（1） 大量使用名词化结构

科技英语要求行文简洁、内容准确，名词化结构是指大量使用名词和名词性词组，将日常用语或其他英语中的动词、形容词等词类充当的某种语法成分转化成名词。例如：

You can rectify this fault if you insert a slash. 在科技英语中转化成：

Rectification of this fault is achieved by insertion of slash.

嵌入一个楔子就可以纠正错误。

The earth rotate on its own axis, which causes the change from day to night. 在科技英语中转化成：

The rotation of the earth on its own axis causes the change from day to night.

地球绕轴自转，引起昼夜的变化。

科技英语文献是以事实为基础记述客观事物的，用词简洁、表达准确，结构严密、描述客观。名词化结构以短语形势来表达一个句子，言简意赅，内部组织密切，应尽量避免使用第一、第二人称。

（2） 广泛使用被动句

科技英语中被动句的大量使用是客观需要的，它注重叙事、推理，强调客观、准确，因而通常采用第三人称来叙述。科技英语中，主语一般位于句首，将“行为”“动作”“作用”“事实”等作为主语，比主动句更加简洁。例如：

The circuit *is broken by* the insulating material.

电路被绝缘材料隔绝了。

Every moment of every day, energy *is being transformed from* one form into another.

每时每刻，能量都在由一种形式转换成另一种形式。

翻译时，要遵循“忠实”和“通顺”两个准则，忠于原文，准确地、完整地、科学地表达出原文的内容，不得删减或者篡改，不得带有感情色彩，使译句通俗易懂。

（3）非限定性动词

在科技英语中，为力求句子简洁，经常用到非限定性动词结构，包括现在分词、过去分词和动词不定式三种形式，可以缩短句子，句意明确。

The parts *to be jointed* are usually heated to a certain temperature.

要连接的这些零件通常要加热到一定的温度。

The type of the manometer depends on the magnitude of pressure *to be measured*.

压力计的形式取决于待测压力的大小。

The ability of a material *to conduct* current depends upon the number of free electrons in the material.

材料的导电能力取决于材料中自由电子的多少。

When two bodies are rubbed together heat is produced, *thus raising the surface temperature of both of them*.

当两个物体摩擦时，就产生了热，因而就提高了这两个物体的表面温度。

The density of air varies directly as pressure, *with temperature being constant*.

若温度不变，则空气密度的变化与压力成正比。

（4）后置定语

在科技英语写作中，大量使用后置定语。长句之所以难以理解，主要是因为后置定语的修饰。常用的句子结构有介词短语后置、定语从句后置、形容词及形容词短语后置和副词后置。这些后置定语放在被修饰词后面，用来修饰这个名词或代词。

1）介词短语后置。

The most widely used methods *for calculating attenuation* are the spectral ratio method, rise time method and the reflection amplitude method.

计算衰减的使用最广的方法是频谱比率法、上升时间法和反射波振幅法。

2）定语从句后置。

The elastic symmetry of a rock is determined in a coordinate system *whose basis coincides with eigenvectors of the averaged acoustic tensor.*

岩石的弹性对称性是在与其平均声张量的特征向量重合的坐标系中测定的。

3）形容词及形容词短语后置。

Seismic tomography is the most powerful method available *for examining the mantle structure*.

地震层析成像是可用来研究地幔构造的最有效的方法。

4）副词后置。

Inorganic flocculants *here* represent nearly 20% of the total flocculants.

无机凝胶在这里代表了总数近 20% 的絮凝剂。

（5）长句

科技英语中常常出现许多长句，习惯用长句表达较为复杂的概念，语法结构比较复杂，从句和修饰语较多。

While the nuclear power, commercial aircraft, chemical, and other industries have taken a conservative approach to introducing new technology, changing designs slowly over time, defense and space systems have pushed the technology envelope, developing tremendously complex, novel designs that stretched the limits of current engineering knowledge, continually introducing new and unproven technology, with limited opportunities to test and learn from extensive experience.

当核能、民航、化工和其他行业都采取保守的方法引进新技术，设计长期处于缓慢变化时，国防和航天系统突破技术界限，研发极其复杂，新颖的设计突破了当前工程技术知识的限制，不断地引进新的尚待检验的新技术，只有有限的机会去测试或从丰富的经验中学习。

FSA can be used as a tool to help in the evaluation of new regulations for maritime safety and protection of the marine environment of in making a comparison between existing and possibly improved regulations, with a view to achieving a balance between the various technical and operational issues, including the human element, and between maritime safety or protection of the marine environment and costs.

FSA可作为一种工具来评价海事安全和海洋环境保护的新规定，对现有规定与可能改进法规进行比较，以期实现各种技术和操作问题之间的平衡，包括人为因素、海上安全或海洋环境保护与成本。

New Progress of Safety Economics

All too often lives are shattered unnecessarily of poor working conditions and inadequate safety systems. Let me encourage everyone to join the International Labour Organization (ILO) in promoting safety and health at work. It is not only sound economic policy; it is a basic human right.

The purpose of this survey is to review the current state of knowledge and issues related to theory and empirical evidence of the market for Occupational Safety and Health (OSH). The increasing competition prompted by globalization, the predominance of service-oriented industries, the rising job insecurity associated with labour market flexibility and demographic developments in the ***composition*** of the workforce, pose important challenges for the health and safety of workers in modern economies. In addition, many governments have recently paid greater attention to the need to tackle the non-trivial costs to both individual and societal welfare that the lack of OSH entails, as part of their overall strive to ***overhaul*** insolvent social security regimes.

An indication of the considerable costs associated with the lack of ***provision*** of OSH is given by estimates of large international bodies such as the World Health Organization (WHO) and the World Bank. They attribute about 3% of lost life years to the factor work. Furthermore, social insurance expenditure on OSH accounts for approximately 2%–3% of GDP in most advanced economies, exceeding by far what is typically spent on unemployment benefits. ILO estimates also show that work-related diseases and accidents account for economic losses as high as 4% of world-wide GDP. Around 4 million accidents at work resulting in more than 3 days of absence occurred in the EU-15 in 2005. This corresponds to an incidence rate of 3100 non-fatal and 3.5 fatal accidents per 100,000 workers. Furthermore, for each worker in the EU-15 an average of 1.3 working days is lost each year due to an accident at work and 2.1 days are lost because of other work-related health problems. A total of 2%–4% of contracted work hours are also estimated to be lost due to sickness absence. Some calculations suggest that the socio-economic costs of absence in advanced Western economies account for between 2% and 3% of their total GDP, that is a typical year's growth.

Similarly, the estimated direct and indirect costs of work-related injuries and illnesses in the USA are approximately $ 170 billion annually. All of the above figures neglect other major non-quantifiable costs, such as private insurance and health care outlays that affected individuals face, the indirect costs that companies incur, the impact on families and communities and the inefficiency of having a large proportion of a potentially active workforce disabled, idle or prematurely retired.

The equilibrium level of OSH in economies is determined by the interplay of incentives faced by workers and firms in labour and insurance markets. This level is influenced by government regulation that is implemented in order to tackle any market inefficiencies that may arise. To the extent that the above costs reflect market inefficiencies, the need for up-to-date information on OSH is of critical importance in order to identify areas of required action and to set priorities for policy initiatives on improving health and safety at work. According to recent ILO estimates, the global number of work-related fatal and non-fatal accidents and diseases has been relatively stable during the past 10 years, despite the significant improvement in OSH attained in many developed economies. This is mainly due to the globalization process and rapid industrialization that have led to a ***deterioration*** of OSH performance in relatively poor countries which are unable to maintain effective OSH standards. Hence the need to focus on health and safety is paramount, given that the traditional hazard and risk prevention and control tools are still effective but need to be completed by strategies designed to address the consequences of a continuous adaptation to a rapidly changing world of work.

Government intervention might be necessary for an economy to attain efficient and equitable levels of OSH. Such government activities typically take the form of the provision of a social security safety net to employees (especially in countries that do not rely on a private insurance market), elimination of informational deficiencies, the setting of standards and the imposition of financial penalties or prosecution to non-compliers. The extent to which such actions are beneficial, though, in the sense that they improve on market outcomes and enhance societal welfare, needs to be assessed rather than assumed.

1. *Social Security*

In most developed countries legally mandated systems exist for providing benefits to workers who suffer from occupational disease or injury. Though in several cases private insurance associations are developed that operate on their own or in parallel with a public social security scheme, it is typical for the state alone to administer compensation for harm in many countries. Adema and Ladaique (2009) report that in the year 2005 approximately 2.6% of the GDP of the OECD group of countries was devoted on average to ***expenditure*** on Incapacity Benefits Significant variations by country are observed mainly due to differences in the criteria for recognition of occupational diseases and eligibility for compensation, with Mexico (0.1%), Turkey (0.2%) and Korea (0.6%) lying at the lower end of the spectrum, the USA (1.3%), the UK (2.4%) and the EU-19 group of countries (3%) found in the middle, and the Scandinavian countries featuring at the top.

Significant cross-country variation also exists in the institutional features governing the provision of occupational disease and injury compensation schemes. In several countries employers are responsible for full coverage of their employees' salary for an initial period (typically 3 days), after which benefits from the social insurance scheme take over. Although the basis for compensation traditionally is to recompense victims for loss of earnings capacity, some countries take into account the damage caused on long-term physical and mental function and diminished quality of life. In most countries a medical assessment of illness or injury is required, which can range from mere validation of disability to the pursuit of a more "aggressive" requirement of job incapacity evaluation. Moreover, there are significant discrepancies in qualifying periods, replacement rates (ranging from 50% in Austria to the total of the ceiling earnings in Luxembourg and Finland), and in the maximum duration of benefit provision.

2. Information Disclosure and Indicators of OSH

When the source of market failure lies in the discrepancy between the perceived and actual risks faced by employees, information disclosure by the government could ameliorate the problem, provided that it is clear, well organized and with sufficiently new information content that avoids information overload. Improving workers' access to information about the nature, severity and relative employment risks that they face is likely to improve market performance, by empowering them to form accurate perceptions of the required wage-risk tradeoff.

Recently, policymakers have sought to construct appropriate measures or indicators of health and safety that will inform the decisions of major stakeholders within their countries. The purpose of such indicators is to quantify the state of affairs in the workplace infrastructure, inputs and outputs of individual economies. As Rantanen et al. suggests, the socio-economic structure of the economy, including demographic conditions, the industrial or service-oriented nature of markets and the technological frontier, are crucial factors for determining the state of workplace inputs and the exposure of different risk groups of employees to them. Exposure to work-related risk factors is, in turn, one of the main determinants of health and safety outcomes, such as the absence behavior of employees, the incidence of work accidents and the occurrence of occupational diseases. These negative OSH outcomes are important as they entail significant direct and indirect economic costs at both an individual and societal level.

3. OSH Regulation, Enforcement and Compliance

There are important differences in the OSH acts and regulations of countries, which usually focus on the criteria required for the receipt and duration of sickness/disability benefits, rules determining the framework under which workers' compensation and tort liability operate and the conduct of OSH inspection and enforcement. The ILO has a central role in providing recommendations and guidance for national OSH policies. Although the ILO publishes instruments, which outline varying levels of obligations for member countries, the latter have to ratify each instrument before implementation takes place. Ratification does not guarantee implementation. The recommendations are for national governments to implement or use as a guiding policy.

A number of studies investigate the factors that underlie the decision to ratify and adopt OSH-related ILO conventions. The length of ILO membership, national income status and regional affiliation are shown to be associated with a higher number of ratifications by member states. Boockmann focus on developing countries, for which OSH is seldom given priority status due to its costly nature. They employ an ***empirical*** methodology that allows for the fact that ratification behaviour is influenced by unobserved characteristics both of countries and of different conventions. The presence of such effects is argued to ***stem from*** the fact that some conventions may be more easily ratified than others, because they differ in terms of their degree of flexibility or complexity. Finally, other factors that are identified as possible obstacles to the ratification process include the lack of infrastructure, the lack of political will, incompatibility with national legal systems and peer effects.

The adoption of government regulations in many countries is often governed by principles that are unrelated to society's WTP (Willingness to Pay) for risk reduction, such as the need to take a precautionary stance against dimly understood risks using an individual risk threshold or a "technology-based" approach. This is ***in contrast to*** the usual economic methodology of assessing the efficiency of regulations by utilizing a cost-benefit approach, where the discounted benefit stream, quantified in terms of the VSL (Value of a Statistical Life), must exceed the cost per life saved. The latter is argued to constitute the most cost-effective method of manipulating the population mean risk level and thus making the best use of society's scarce resources, although important objections are expressed on the grounds that cost-benefit analysis is unreliable and unethical. Furthermore, it is pointed out that significant caution needs to be exercised when undertaking stringent regulatory actions, given that offsetting behavioural consequences (moral hazard) may negate any beneficial impacts of the regulation. For instance, the positive effect of any legislation forcing employers to supply personal protective equipment to their employees might be mitigated by the subsequent complacency of workers, thus leading to more workplace accidents. This is particularly likely to be the case when people overestimate the efficacy of a policy. The overall assessment of the ***efficacy*** of regulations is also complicated by the fact that there might be substantial opportunity costs associated with the diversion of resources away from other potentially health-enhancing expenditures.

The success of OSH regulation is also dependent on enforcement by the relevant authorities of the stipulated OSH standards and compliance on behalf of the affected agents. Because compliance with a regulation sometimes involves substantial administrative and other costs, rational firms will choose to comply provided that the expected costs of noncompliance are greater. Given that the latter are a function of the likelihood of detection and the ***magnitude*** of the expected penalty levied, it follows that a low probability of detection plus low fines for violations of OSH standards are likely to compromise the efficiency of regulation. Empirical evidence tends to suggest that the estimated effects of OSH inspections on safety are quite small or non-existent, although research in this area has found it difficult to circumvent the problem of the endogenous nature of injuries and inspection.

New Words and Expressions

all too often　　时常，经常是
composition [ˌkɒmpəˈzɪʃn]　　*n.* 成分，作文，构成，创作
overhaul [əʊvəˈhɔːl]　　*v.* 彻底检修，赶上
provision [prəˈvɪʒ(ə)n]　　*n.* 提供，条款，供给，供应品
deterioration [dɪˌtɪərɪəˈreɪʃn]　　*n.* 恶化，退化，变质，堕落
expenditure [ɪkˈspendɪtʃə]　　*n.* 开支，费用，消费，消耗
empirical [emˈpɪrɪk(ə)l]　　*adj.* 凭经验的
stem from　　起源于
in contrast to　　与……形成对照
efficacy [ˈefɪkəsi]　　*n.* 效力
magnitude [ˈmægnɪtjuːd]　　*n.* 震级，巨大，重大，重要性

Unit Three

Statistics for the Safety Professional

Statistics are extremely important in the career of the safety professional. We utilize them to report our annual injury and accident rates, to calculate failure probability rates, and to describe observed or experimental data. For our purposes, the definition of statistics is the study of obtaining meaningful information by analyzing data. This data can come from two different sources. It can come from experiments in a controlled environment or from observations made. Statistics can reduce large sums of data to a ***manageable*** form, and allow the study and analysis of variance, thus allowing managers to maximize the use of information available to form an effective decision. The study and instruction of statistics can ***take up*** volumes of text. We will begin the review of statistics by describing descriptive ***statistics***.

1. Descriptive Statistics

Descriptive statistics are just what the name implies. They are used to describe a set of data. For example, we have all heard polls on the local news that describes the data as 52% of all Americans are opposed to the latest congressional bill plus or minus 3%. What this is stating is that the range of the population that disapproval is 55% to 49%. This is a measure of central tendency.

2. Mean

The mean is also referred to as the "average" or arithmetic mean. For a data set, the mean is the sum of the observations divided by the number of observations. The mean is often quoted along with the standard deviation. The mean describes the central location of the data, and the standard deviation describes the spread. The mean is represented by the Greek letter (μ), which is *mu*, "pronounced moo". It can also be represented by *x*. The mathematic representation for calculating the mean is as follows

$$\mu = \bar{x} = \frac{x_1 + x_2 + \cdots + x_n}{n}$$

where μ and $\bar{x}$ = average or arithmetic mean;

x_n = individual data values.

3. Mode

The mode of a data sample is the variable that occurs most often in the collection. For example, the mode of the sample {1, 3, 6, 6, 6, 6, 7, 7, 12, 12, 17} is 6, since it occurs 4 times. To find the median, your numbers have to be listed in numerical order, so you may have to rewrite your list first.

4. Median

The median is the middle value in the list of data. To find the median, your numbers have to be listed in numerical order, so you may have to rewrite your list first. If there is no "middle" number, because there is an even set of numbers, then the median is the mean (the usual average) of the middle two values. For example, given the data set {1, 2, 3, 4, 5, 6, 7, 8}, what is the median? The median would lie between 4 and 5, therefore add 4 + 5 = 9, then divide 9 by 2, giving a product of 4.5. Therefore the median for the data set is 4.5, which is the central point of the data set.

5. Variance

The variance can be described as the degree to which the variables in the data set are ***spread out***. In other words, we would like to know how far away from the mean the variables are. To determine the variance from the mean, we would calculate each data point and the mean of the data set. For example, let us assume that you are the corporate safety manager for a corporation having six different and ***distinct*** locations.

To utilize this data in a meaningful way, we cannot simply add the spread and divide by the number of points. By doing this the positive and negative numbers would simply cancel each other out and be equal to or near zero. Therefore, it is necessary that we square the distances from the mean (average). To do this we utilize the following equation

$$\mathrm{Var}(x) = \frac{(x_1 - \bar{x})^2 + (x_2 - \bar{x})^2 + \cdots + (x_n - \bar{x})^2}{n}$$

Or written another way

$$\mathrm{Var}(x) = \sigma^2 \frac{\sum_{i=1}^{n} (x_i - \bar{x})^2}{n}$$

The variance is represented by the Greek symbol σ^2 and is read as sigma squared.

6. Normal Distribution

Let's now discuss normal distribution. Most of us are familiar with the "bell curve". The bell curve is basically a graph of normal distribution, which has a single peak and demonstrates that half of the data points are on the left side and half of the data points are on the right side of the curve. The mean lies in the center. The two tails extend indefinitely, never touching the horizontal axis. No matter what the value of the mean and the standard deviation, the area under the curve is 1.00. The variance and standard deviation are measures of the variability of a data set with respect to its mean. The graphic representation is shown in Figure 3-1.

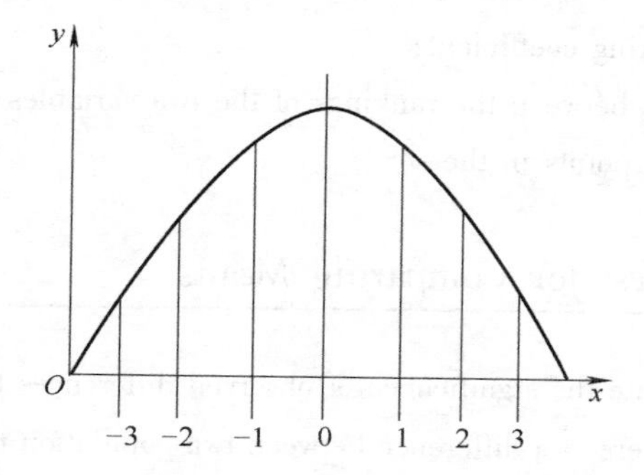

Figure 3-1 Graphic Representation of Standard Deviation

As mentioned previously, the entire bell curve value is 1.00, regardless of the value of the data points. The numbers along the x axis of this graph represent the number of standard deviations in the distribution. Standard deviation values are listed below that you should commit to memory, as there may be several questions on the examination related to standard deviations:

1 std. dev. = 68% of the data set

2 std. dev. = 95.45% of the data set

3 std. dev. = 99.73% of the data set

7. Calculating Correlation Coefficient

The correlation ***coefficient*** is also known as the product moment coefficient or Pearson's correlation, and is described in the following equation

$$\gamma = \frac{n\sum(xy) - (\sum x)(\sum y)}{\sqrt{[n\sum(x^2) - (\sum x)^2][n\sum(y^2) - (\sum y)^2]}}$$

where γ = sample correlation coefficient;

n = number of data points;

$x = x - \bar{x}$;

$y = y - \bar{y}$.

This equation is used to determine if there is a correlation between variables. It does not necessarily mean that there is a causation. It simply means, there is a strong possibility that if x is this, then y can be predicted to be that, similar to regression analysis. It indicates the strength of a linear relationship between variables.

8. Spearman Rank Coefficient of Correlation

One method of testing a hypothesis of a correlation between variables is to use the Spearman rank coefficient. The equation is shown mathematically as

$$\gamma_s = 1 - \frac{6\sum(D^2)}{n(n^2-1)}$$

where γ_s = Spearman ranking coefficient;

$D = \Delta$ or difference between the rankings of the two variables;

n = number of data points in the set.

9. Calculating the t-Test for Comparing Means

The t-test is used to calculate the significance of observed differences between the means of two samples. It can be used to determine if there is a difference between two population parameters. In addition, it can be used in situations where one of the conditions, the population standard deviation, is not known and the sample size is ≤30. It is generally used with scalar variables, such as length and width. The null hypothesis is that there are no significant differences between the means. To do this we use the following equation

$$t = \frac{\bar{x}-\mu}{s}\sqrt{n-1} = \frac{\bar{x}-\mu}{\hat{s}}$$

10. Chi-Square Statistic

The chi-square (χ^2) statistic is useful in comparing observed distributions to theoretical ones. To calculate the chi-square (χ^2) test, we use the following equation

$$\chi^2 = \sum_{j=1}^{k}\frac{(o_j - e_j)}{e_j}$$

where o = observed data;

e = expected data.

11. Permutations and Combinations

There is no easy definition of ***permutation*** or combination, in regards to mathematics. Therefore, it is necessary that we explain the difference between the two. To illustrate what each are, an example of each is listed below:

Combination: A group of data or items where an order doesn't matter. For example, your meal ***consists of*** chicken, salad, bread, and vegetable. Bread, Salad, vegetable, and chicken is still your meal. In other words, no matter what order you place the items in, it is still the original meal.

Permutation: A group of data or items where an order does matter. For example, the numbers to a combination lock are 5-8-6, which is not the same as 8-6-5. In this example, the order matters. A permutation is an ordered combination.

Note: If the order doesn't matter, it is a combination. If the order matters, it is a permutation.

There are two types of permutations. The first type is when ***repetition*** is allowed. For example, the lock mentioned above may have a data set of 4-4-4. The second type is when no repetition is allowed, as in describing the first three in an automobile race.

12. Permutations with Repetition

To calculate permutations with repetition, simply multiply the data points as described in the equation below:

$$n^r = n \times n \times \cdots \times n$$

where n^r = permutation when repetition is allowed;

n = the number of possibilities to choose from.

In this example there are ten possibilities to choose (0, 1, 2, 3, …, 9) from and you are selecting three of them, therefore the permutation for this data set is solved mathematically as follows

$$n^r = 10 \times 10 \times 10 = 10^3 = 1000\text{permutations}$$

As you can see, calculating permutations with repetition is quite simple.

13. Permutations without Repetition

Calculating permutations with repetition is slightly more complicated. To calculate the permutation for the number of the combination lock mentioned previously where the numbers are 5-8-6, which is required to be in a specific order, we would use the following equation

$$p_k^n = \frac{n!}{(n-k)!}$$

where p_k^n = permutation;

n = number of data points to choose from;

k = number of data points you choose.

New Words and Expressions

manageable [ˈmænɪdʒəb (ə) l] *adj.* 可操纵的，可处理的

take up 拿起；开始从事；占据（时间、地方）

statistics [stəˈtɪstɪks] *n.* 统计学，统计数字

spread out 展开，铺开，伸张

distinct [dɪˈstɪŋ (k) t] *adj.* 清晰的，清楚的，明白的，明显的

coefficient [ˌkəʊɪˈfɪʃ (ə) nt] *n.* 系数，(测定物质某种特性的) 系数

permutation [pɜːmjʊˈteɪʃ (ə) n] *n.* 置换，排列（方式），组合（方式）

consist of 包含，由……组成，充斥着

repetition [repɪˈtɪʃ (ə) n] *n.* 重复，重做，重说，重做的事

Exercises

1. What is statistics? What is safety statistics?
2. Describe the variance. How to determine the variance from the mean?
3. Describe the normal distribution and explain why the bell curve is basically a graph of it.
4. Explain the difference between the permutation and combination.
5. Give an example of the difference between the permutations with repetition and permutations without repetition.

科技英语翻译技巧——方法简介

科技英语的翻译技巧有直译法和意译法、增译法和省译法、合译法和拆译法、顺译法和倒译法等。

1. 直译法和意译法

直译法是在保留原文的语法和结构的基础上，忠实地表达原文的内容与形式的一种翻译方法。意译法是指不拘于原文的形式，将原文的内容表达清楚即可，可适当地省略、增补、词义转换和引申，也可改变句子的结构和表达方式的一种翻译方法。在科技英语翻译中，应尽量采用直译法，在直译法无法清楚地表达原文的意思时，可选择意译法。

（1）直译法

直译法在科技英语中使用较多。例如：

Oil and gas will continue to be our chief source of fuel.

石油和天然气将继续是燃料的主要来源。

Physics studies force, motion, heat, light, sound, electricity, magnetism, radiation, and atomic structure.

物理学研究力、运动、热、光、声音、电、磁、辐射和原子结构。

（2）意译法

在科技英语中，意译强调的是“神似”。例如：

We can get more current from cells connected in parallel.

电池并联时提供的电流更大。

Resistors are available either in fixed values or variable values.

电阻器有固定和可变两种。

2. 增译法和省译法

（1）增译法

在翻译时增加原文省略的内容或者原文中有其含义的词语，从而满足既能准确地表达原文的含义，又能符合汉语的表达习惯的修辞需要。例如：

High technology is providing visually and hearing impaired people with increased self-sufficiency.

高科技设备在不断地增强视力和听力损伤者的自理能力。（增译名词“设备”和副词“不断地”）

（2）省译法

由于英语和汉语的表达方式不同，为使译文准确地表达出原文的思想内容，有时需要将一些词语省略，例如：

For the purpose of our discussion, let us neglect the friction.

为了便于讨论，我们将摩擦力忽略不计。（省译名词“purpose”）

3. 合译法和拆译法

合译法和拆译法是在句子过于简单或者过于烦琐时采取的翻译方法。

（1）合译法

合译法就是在翻译时，把原文中表达内容相同的两个及两个以上的简单句或者复合句，翻译成一个句子。

There are some metals which possess the power to conduct electricity and ability to be magnetized.

某些金属具有导电和被磁化的能力。

（2）拆译法

拆译法是指把长句中的从句或短语化为句子，分开来叙述。例如：

It is very important that a machine element be made of a material that has properties suitable for the condition of service as it is for the loads and stresses to accurately determined.

载荷和应力必须准确地计算出来，机器零件要用性能符合工作条件的材料来制造。这两件事都是非常重要的。

4. 顺译法和倒译法

（1）顺译法

当原文是按时间或逻辑顺序叙述，且与汉语的表达方式相同时，采用顺译法。例如：

Moving around the nucleus are extremely tiny particles, called electrons, which revolve around the nucleus in much the same way as the nine planets do around the sun.

围绕着原子核运动的是一些极其微小的粒子，称为电子，这些电子围绕着原子核旋转，像九大行星围绕太阳旋转一样。

（2）倒译法

当英语长句的顺序与汉语表达方式不一致时，常常使用转换、颠倒、改变部分或完全改变词序的倒译方法。例如：

It may be economically sound, in the long run, to subsidize their initial production, event at prices above the projected marked for natural hydrocarbon fluids, in order to accelerate the deduction of dependence oil imports.

从长远的观点来看，资助开发气体等燃料，即使价格高于自然碳氢化合液的市场价格，但为了加快减少对进口石油的依赖，这在经济上可能还是合算的。

Explorative Spatial Analysis of Traffic Accident Statistics and Road Mortality among the Provinces of Turkey

Road traffic accidents are increasingly being recognized as a growing public health problem in developing countries. It is estimated that in the European Union 1. 3 million traffic accidents occur every year, resulting in over 40,000 deaths. Turkey, a rapidly developing country, is a junction point between Asia and Europe in terms of its social and economic structure. The population is 70, 586, 256, and the number of drivers license holders is 18, 877, 354 according to the 2007 population census. There is an explosion in immigration and population, and a corresponding increase in ***vehicle*** numbers. Meanwhile gross domestic product and income per capital have grown rapidly in recent years. The increase in motor-vehicle ownership is very high in Turkey, 795, 552 new vehicles being registered in 2007 alone. Although the population increased by

14.70%, motor-vehicle ownership increased by more than 75% between 1995 and 2007. During that period, the number of traffic accidents and injuries also increased. The number of deaths, however, decreased by 42.39%. The amount of material loss owing to traffic accidents was approximately $ 1.1 billion in 2007, excluding health expenses and loss of labor.

Previous statistics have shown that ***casualty*** and fatality rates in Turkey are much higher than in developed countries with comparable vehicle ownership levels. Compared with other countries, Turkey's traffic death rates are ***deplorable***. The rate of fatal accidents in Turkey is 20 per 100 billion vehicle-kilometers, whereas the rates in the UK, United States, and Germany are 0.9, 1.1, and 1.6 respectively. In the last decade, the rate of accident per 100 million vehicle-kilometers more than doubled. Turkey, having fewer vehicles than other countries, is recognized as being unlucky in terms of fatal accident numbers.

Rapid expansion of road construction and increased numbers of vehicles means that road traffic accidents are becoming an increasingly serious public health problem in Turkey. Achieving ***reductions*** in the number of traffic accidents and deaths is a national priority. Despite the significant impact of traffic accidents and mortality on public health, the magnitude of the effects could be greatly reduced if preventative measures were taken in the safety-deficient provinces. The identification of safety-deficient regions on the highway network is aimed at comprehensive safety programs. ***Geographical Information Systems*** (***GIS***) are a very important and comprehensive management tool for traffic safety. GIS-aided spatial analysis provides information on hazardous regions, hot spots, warm spots, and so forth. The spatial component of traffic accidents has always interested researchers and GIS-users. Empirical examples of spatial analysis of traffic accidents are mainly concerned with black spot analysis in local areas, with a few exceptions ***regional*** data are often neglected. Regional variations in traffic accidents and mortality in Turkey have also received no attention so far.

This article examines the regional disparities hidden behind national statistics on road accidents and fatalities in Turkey using GIS and ***spatial analyses***. It is a common practice to compare cities or countries in terms of road safety performance and to rank them in terms of risk indicators such as the accident or death rates, which are often expressed as the number of accidents/deaths per 100,000 inhabitants. In Turkey, regional differences in traffic accidents and traffic accident mortality and their underlying determinants have not been studied, although Turkey has large regional differences in socio-economic development. The study therefore describes preliminary investigations of road accidents and mortality in Turkey at the province level, adjusting for population and number of registered motor vehicles owing to socio-economical differences. The number of accidents and mortality ratio are modeled with variables through geographically weighted regression. Different types of software were used for visualization and spatial analyses of the accident data in the study, developed by Environmental Systems Research Institute (ESRI), GeoDa 0.9.5-I developed by Luc Anselin through the Center for Spatially Integrated Social Science at the University of Illinois, CrimeStat 3.1 developed by Ned Levine, with support from the National Institute of Justice and GWR 3.0 developed by Fotheringham, Brunsdon and Charlton.

Accident reports are prepared by police departments within the borders of municipalities and by the gendarmerie headquarters in the provinces without police responsibility. Spatial analyses were performed on the aggregated data reported by both police and gendarmerie in the provinces of Turkey from 2001 to 2006. Population by census year, annual intercensal rate of increase and mid-year population forecasts, number of traffic accidents, and records of road deaths in the provinces were taken from the data held by the Turkish

Statistical Institute. Socio-economic index values of the provinces, calculated by principal component analysis, including 58 factors regarding demographic, education, employment, health, industrial, economic, agricultural, and construction status were taken from the State Planning Agency (SPA) for 2003.

Analyses in this study are based on the province levels of spatial ***aggregation***. Aggregated area-based data are very important sources of information for many social science disciplines. Geographical locations of these data are also an important factor in many aspects of social and economic policies at the national level because demographic, social, and economic characteristics of society represented by aggregated province units show differences throughout the country. The province unit is the common level for social, economic, demographic, and administrative data collection by agencies in Turkey. Usage of province units in analyses therefore allows comparison of accident and mortality rates with other demographic and economic features. Accident and death rates were examined with spatial analyses at the province level in this study. Province units, however, have important limitations; provinces are administrative units, and cover large areas with different ***heterogeneous*** populations, and they may not match the ecological scale. It has been suggested that aggregating the accident and mortality rates for the entire 5-year period provides the advantage of stability in the province-level accident and death rates, and also summarizes the phenomenon. A temporal equilibrium state was therefore examined by means of the time series of mortality and number of accident cases, and 2001–2006 emerged as a relatively stable period.

Population density and the number of registered motor vehicles were used as standardization factors in this study. The Mortality Rate (MR) standardized with population is the number of deaths based on traffic accidents in a province during one year divided by the total number of inhabitants residing in that province in the middle of that year:

MR_ P (standardized by population)

= 100, 000 Number of deaths/Total number of inhabitants.

The MR standardized with numbers of registered motor-vehicles is the number of deaths based on traffic accidents in a province during one year divided by the total number of registered motor-vehicles in that province in the middle of that year:

MR_ M (standardized by number of registered motor-vehicles)

= 100, 000 Number of deaths/Number of registered motor À vehicles.

The Accident Rate (AR) standardized by population is the number of accidents resulting in death or injury in a province during one year divided by the total number of inhabitants residing in that province in the middle of that year:

AR_ P (standardized by population) = 100, 000 Number of accidents resulting in death or injury/ Total number of inhabitants.

The accident rate standardized by number of registered motor-vehicles is the number of accidents resulting in death or injury in a province during one year divided by the total number of registered motor-vehicles in that province in the middle of that year:

AR_ M (standardized by number of registered motor-vehicles)

= 100, 000 Number of accidents resulting in death or injury/Total number of registered motor À vehicles.

These are not true rates, as the populations in the provinces are not equal to the populations that are exposed to the risk of road death. Some inhabitants reside in the province but work or travel in another province, and some motor-vehicles are registered in the province but are used for work in or travel to another province. Hence, this index does not consider mobility. It uses the simplest and most available variable as the ***denominator***: Total resident population and total number of registered motor-vehicles, which are currently the most appropriate variables for ***comparative*** purposes.

A commonly-used concept in rate analysis is that of a standardized mortality rate, or the ratio of the observed mortality rate to a national standard. The excess risk is the ratio of the observed rate to the average rate computed for both the number of accidents and mortality. This average is not the average of the provincial rates, but calculated as the ratio of the total sum of all cases over the total sum of all populations at risk. An excess risk rate greater than 1.0 indicates that more fatal accidents occurred than would have been expected; a ratio of less than 1.0 indicates fewer deaths than expected.

Since the accident data were aggregated into the areal units of provinces, an important aspect is the ***derivation*** of a spatial weight matrix (W) for spatial analyses. W is the fundamental tool used to model the spatial proximity and interdependence between areal units. Determination of the proper W matrix is a difficult and controversial topic in spatial analyses. There are wide ranges of W determination applications according to the aim of the studies covered in the literature. In this study, three different methods were used to obtain W matrices. The first and second matrices were calculated by the criterion of ***contiguity*** according to the centroid of the nearest 5-10 neighbors. The third matrix was formed according to the criterion of general social distance.

When aggregated data are in use, if the population or the number of cases is relatively small or sparse, rate estimates may not be precise. In order to overcome the problem of rate instability, various smoothing methods are usually employed. The idea of smoothing is to borrow information from other small areas for the estimation of the relative risk. In this study, Empirical Bayes (EB) smoothing was used and raw rates were replaced by their globally-smoothed values calculated by an EB tool in ArcGIS 9.2 (National Cancer Institute of USA). Smoothed rates can be mapped to identify more clearly some structural forms different from those induced by density. Choropleth maps provide an effective way of visualizing how rates vary across a region. Smoothed rates provide accurate visual representation of the standardized rates compared with the raw mortality and accident rates.

New Words and Expressions

vehicle [ˈvɪəkl] *n.* 车辆，工具，交通工具

casualty [ˈkæʒjʊəlti] *n.* 受害者，遇难者，毁坏物，损坏物

deplorable [dɪˈplɔːrəb (ə) l] *adj.* 糟透的，令人震惊的，令人愤慨的

reduction [rɪˈdʌkʃ (ə) n] *n.* 减少，降级，(刑罚等的) 减轻，减速

Geographical Information Systems (GIS) 地理信息系统

regional [ˈriːdʒənl] *adj.* 地区的，区域的，地方的

spatial analysis 空间分析

aggregation [ˌægrɪˈgeɪʃən] *n.* 聚合，集团，群聚，族聚

heterogeneous [ˌhet (ə) rə (ʊ) ˈdʒiːniəs] *adj.* 由很多种类组成的，各种各样的

denominator [dɪˈnɒmɪneɪtə] *n.* 分母

comparative [kəmˈpærətɪv] *adj.* 比较的，相比的，比较而言的，相对的

derivation [derɪˈveɪʃ (ə) n] *n.* 推导，衍生，导出

contiguity [ˌkɒntɪˈgjuːəti] *n.* 接近，一连串的事物，一系列，一大片

Unit Four

4

Ergonomics

Some men were at work shifting sea containers in Malaysia. After fitting the twist lock to a container before it was lifted one of the men disappeared from view. The crane on the quay moved, the man was dragged over six metres and died at the scene. A ***subsequent investigation*** reported that there was a high probability that he had been resting against the wheel of the crane.

The investigation also found that there were no rest areas and no guidance for employees. The recommendations included stricter supervisory control, a warning buzzer on the quay crane, warning signs, and an area where a worker could rest in comfort and safety. This is of particular importance for workers involved in ***manual*** labour carried out in a hot and/or humid environment.

1. What Is Ergonomics?

Ergonomics comes from the Greek words "ergon" (work) and "nemein" (to arrange or manage). Today we understand it to mean a study of the fit between a person and the elements of the task they are required to perform. This is not confined to the workplace—it could, for example, apply to any car driver. In the United States this study has, until recently, been described as "human factors engineering". Ergonomics aims to design the task for the person, not the other way round. The ancient Greek story of Procrustes illustrates the point. Procrustes used to cut a visitor to fit the bed he provided, not adjust the bed. In ergonomics the layout, management system, methods, equipment and environment are all considered in relation to the inherent capabilities and limitations of human beings.

2. Research and Development Work in the United Kingdom, America and Australia

Some early work on ergonomics was done by a Polish researcher, Jastrzebowski, in the nineteenth century. But even prior to this some of the principles were recognized by Bernardino Ramazzini when he wrote his book *On the Diseases of Occupations* in 1700. The term "ergonomics" was coined by Jastrzebowski in 1857. It was reintroduced in the UK in 1950, but the study of the person within the system in the United States, known as human factors, which has a variation of emphasis, goes back well beyond that.

The heat stress to which submarine and tank crews were subjected was studied during the Second World War. Work and physiological demands were also considered. Efficient armaments production in World War Ⅰ, when large numbers of extra women became part of the workforce, demanded attention to methods of designing production. Working hours and overtime in relation to production, and relationship of temperature to accidents, were studied. The interpretation of radar and sonar signals by the operators involved new ergonomic considerations, because the senses were extended. Extending the sight sense with infra-red night goggles deserves care, as the 1996 Blackhawk helicopter disaster in the Australian Army showed.

Work-study, including time and motion studies, looked at efficient ways to design production. Tayloristic production-line methods paid too little attention to the whole person and their needs and, although they are now recognized as a mistake, they may have been the best approach in, for example, the World War II production of Liberty ships.

Studies of human performance have been many and varied. They include human response to controls (dials, levers, switches, displays), human performance under high and low gravity (for fighter aircraft and orbiting space stations), human behaviour in relation to complex systems (nuclear power station operation), and effective sorting (mail exchanges). More recently the interaction of humans and ***computer terminals*** has received a lot of attention.

Matching the physical dimensions of a workstation with the dynamic and static physical dimensions of people is known as anthropometry. Anthropometry has been used to design many things for effective and compatible use by the majority of people—for example, bench and chair heights—but it is only relatively recently that children have had purpose-designed toilet pedestals, for example. Biomechanics has been used to study the muscles, tendons, ligaments, joints and bones used in lifting, pushing, pulling and twisting. Anthropometry and biomechanics have been utilized together to examine the design of, for example, hand tools.

Heat stress has also been studied in army personnel undergoing rigorous training, initially by Yaglou and Minard in the USA (and continuing to this day at Fort Detrick); in UK submarine crews by Bedford; and in bushfire fighters in Australia by Brotherhood and Budd. Performance in Antarctic cold has been studied by the latter researchers and United States researchers such as Siple and Passel.

Shift work, its psychological effects and the fatigue associated with it, have also been the subject of intensive study by researchers such as Moore-Ede in the USA, and Wallace in Australia. ***Repetitive*** tasks such as writer's ***cramp*** have been reported on since a late nineteenth century study of people working all day with quill pens.

Clearly, the needs of physically or mentally challenged people form an area of special consideration in ergonomics—for example, rehabilitee, and those in "sheltered workshops". A visit to a centre displaying items for the challenged reveals the thought which has gone into overcoming the problems associated with reduced sight, reduced movement, reduced hearing, reduced motor coordination and so on.

3. Scope and Levels of Ergonomic Activity

Three different levels of work which involve ergonomics can be found in the workplace. They are: work station design, workplace design, and job design.

Work station design involves the disciplines mentioned above—anthropometry and biomechanics (which involves functional anatomy and physics) —as well as engineering and psychology. Psychology is involved because it ***looks at*** how people receive information or ***stimuli***; how they process this, or these, in the brain; how they make decisions; and then how they act upon them. Experience and training can improve perception and decision making. Perception of some stimuli, such as visual ones, is affected by old age.

Workplace design involves fitting work stations into the overall physical design of a workplace. This will include issues such as noise, temperature, lighting, colour and rational movement of people and materials.

Job design takes into account what has been said above, but also looks at the way tasks are broken up, the decision-making processes and the operations of work groups. Important considerations are level of mental stimulation (under or over); conflict between different people and different work areas; meaning in the work; and degree of control (consultation). Motivation and hence safety culture are important elements in this. The psychology of errors and types of errors are important aspects of safety. Physiology plays a part in examining the way people perform, particularly in extremes of temperature and humidity.

4. Development of Ergonomics for the Work Station, Workplace and Job Design Levels

The earlier description of some areas of ergonomics in which research and development has taken place can be analysed further using the levels just identified; those are work station, workplace and job design. Clearly many areas of what we consider mainstream "workplace safety" fit these ergonomic levels. Think of, for example, machine guarding using mesh to prevent hand entry; pedestrian zone markings in workshops and safe job procedures; and the maintenance of vigilance and attention to detail, which can be affected by, for example, heat, noise and vibration. So some of the trends developed over time in the safety arena can now be seen just as surely as ergonomic advances.

5. Integration of Human, Machine and Environmental Factors in Work Stations and Workplaces

There are a large number of examples of work station and workplace scenarios which illustrate the integration of human, machine and environmental factors. Three examples of those will be given here:

Pliers, paint scrapers and the pistol-grips on electric drills all require attention to the tendons, muscles and joints of the hand and arm used to grip and control the device. A close examination is needed to ensure that the position of the parts of the hand doesn't result in early ***fatigue*** and cramp, and inappropriate direction of application of force through the forearms. Grip will be affected by sweating due to environment and physiological (work effort) factors. Chain saws and building site soil compactors add vibration to this list of environmental factors.

Control of an aircraft involves integration of a complex range of information about the machine and its spatial position in the environment. The vertical angle in relation to the line of flight governs stall; lift is affected by speed and flaps; and information about height and banking angle may be partly visual and partly through pressure receptors in the skin, but also relies on instruments. The psychology of ***perception*** becomes important. The Air New Zealand aircraft disaster on Antarctica's Mt Erebus in 1986 resulted from an inability to distinguish sky from landscape in certain weather conditions (plus an error in programming the flight control computer). Wind shear on landing requires split-second ability to respond.

Assembly and soldering of coloured wiring or connections to printed circuit boards with further assembly, requires the ability to distinguish colours clearly—remembering that a significant percentage of men, in particular, are colour-blind—and to maintain care and ***vigilance*** during repetitive work. Noise and heat or humidity will affect concentration, but heat and humidity are often controlled not so much for the operator's comfort as to protect the electronics.

6. Ergonomic Principles of Work Stations

Working posture in general can be considered in terms of task requirements, workspace design and a range of personal factors. Enforced sitting for most of a work day should be considered an occupational risk factor.

The key musculoskeletal issues involve consideration of:

- the position of the spine;
- the effect on the discs between the vertebrae;
- the effect of, and on, posture;
- fatigue associated with static muscle loading;
- the effect on muscles of reduced blood supply;
- the effects of restricted movement [enforced by, for example, the need to view a VDU (Visual Display Unit)] on the head and neck;
- the different muscle groups which are used for lifting while seated because the full set of muscles in the body cannot be employed.

Progressive or even acute spinal damage can occur from jarring while seated at the work station in off-terrain vehicles. Persistent pressure on discs due to spinal position while seated can be a source of fatigue and discomfort. Measurements have been made and show the least pressure on the disc between the third and fourth lumbar (lower back) vertebrae when a person is leaning backwards. The pressure increases in the writing position, even more when typing and more again when lifting a weight from the desk or table. When sitting, the upper edge of the pelvis tilts backwards and changes the bow (from bow and arrow) shape of the spine, called the lordosis, into a backward curve (kyphosis). People find a slight bend forwards comfortable because it requires less muscle strain. There is a conflict between the muscles, which prefer the curvature, and the discs, which require an upright posture.

Discs require nourishment, and this is achieved in part by frequent changes of pressure—such as can occur when posture is adjusted. Further work has shown that the best angle between thigh and back is 120 degrees. To make work at a desk feasible while retaining this angle, the knee chair has been introduced.

Fixed posture increases the static loading (the work muscles must do to hold the body in one place) on the back and shoulder muscles. It can also reduce blood flow to the legs and so lead to swelling and discomfort.

The following recommendations are made in relation to seating:

- Promote lumbar lordosis (the bow shape in the spine) —a lumbar support between seat back and spine helps.
- Minimize pressure on discs. Reclined backrests, lumbar support and arm rests all help; arm rests reduce the work shoulder muscles need to do.

• Minimize the static loading on the back muscles. An angle between squab and backrest of around 110 degrees helps.

• Reduce fixed posture. Take breaks away from, for example, VDUs and VDU work if possible.

• Use chairs which can be readily adjusted, and show employees how to use them: seat height and slope, depth and width, contouring and cushioning and seat-back design all need to be considered. Different people prefer different seat designs.

New Words and Expressions

subsequent investigation	事后调查
manual [ˈmænjʊ(ə)l]	n. 说明书，指南，使用手册
computer terminals	计算机终端机，计算机终端设备
repetitive [rɪˈpetɪtɪv]	adj. 重复的，啰唆的，复唱的
cramp [kræmp]	n. 抽筋，痛性痉挛，（腹部）绞痛
look at	研究
stimuli [ˈstɪmjʊlaɪ]	n. 刺激，刺激物，促进因素
fatigue [fəˈtiːg]	n. 疲劳，疲乏，杂役 v. 使疲劳，使心智衰弱，疲劳
perception [pəˈsepʃ(ə)n]	n. 感知，知觉，看法，洞察力
vigilance [ˈvɪdʒɪl(ə)ns]	n. 警惕，警戒，[医] 失眠症

Exercises

1. Explain the definition of ergonomics.
2. What aspects does ergonomics involve?
3. What three levels of work which involve ergonomics can be found in the workplace?
4. What aspects should we consider when designing a work place?
5. Give an example of work station and workplace scenarios which illustrate the integration of human, machine and environmental factors.

科技英语翻译技巧——词量的增加

由于英语和汉语的表达方式不同，在科技英语翻译中，常常会增加一些词语来使表达更为准确。更通顺易懂。增加的词必须是文中没有出现，但是意义上存在的一些词，使得译文在语法和语言上更符合汉语的习惯。

增词可遵循以下原则：

1）原文不曾出现的词语，但句意存在其意思，需要增加词语。

2）增加的词语起到辅助功能，帮助句子更加通顺、符合汉语的习惯，但不能增加或者改变原句的意思。

3）增词的目的在于使译句更加通顺，更能表达原句的准确意思，不能画蛇添足。

1. 增加表示名词复数概念的词语

在翻译时，遇到名词的复数形式时，可根据具体的情况，适当增加表示复数概念的词，如“一些”“们”“许多”“之类”“大量”“几个”“几次”等。

Carbon combines with oxygen to form carbon oxides.

碳同氧化合形成多种氧化碳。(增译“多种”)

In spite of the difficulties, our task was got over well.

尽管有各种困难，我们仍然顺利地完成了任务。(增译“各种”)

Solid in greatly different degrees resist being changed in shape.

各种固体抗形变的程度极不相同。(增译“各种”)

2. 增加原文省略的词语

在英语句子中，会省略某些已经在前文出现过的词语，使句子简单不啰嗦，因此，在翻译时需要补充这些省略了的词语，保证句子的完整性。

When a material is subjected to a tensile or compressive load of sufficient magnitude, it will deform at first elastically and then plastically.

当材料受到一定大小的拉伸荷载或压缩荷载作用时，它首先发生弹性变形，然后发生塑性变形。(增译“发生”)

The changes in matter around us are of two types, physical and chemical.

我们周围的物质变化有两种，物理变化与化学变化。(增译“变化”)

High temperatures and pressures changed the organic materials into coal, petroleum and natural gas.

高温和高压把这些有机物变成了煤、石油和天然气。(增译 pressure 前的“高”)

3. 增加具体化、明确化的词语

Combine digital technology with advanced software, smaller and more powerful microprocessors, and exponential growth in fiber and wireless bandwidth, and you get something far more powerful-seamless, universal connectivity.

把数字技术与先进的软件，体积更小、功能更强大的微型处理器，以及呈指数增长的光纤和无线带宽相结合，你就会获得功能更强大的无缝的全方位的连接。(增译“体积”“功能”)

Gravity is a strange force, when you slip on something, you never go into the air, but instead you always fall down.

地心引力是一种奇异的力量，当你滑倒时，你就不会跌向空中，而是相反，总是倒向地面。(增译“地面”)

Were there no electric pressure in a conductor, the electron flow would not take place in it.

导体内如果没有电压，便不会产生电子流动现象。(增译“如果”)

4. 增加必要的连接词

为了使译文通顺，使句子与句子之间更加连贯，关系更加紧密，因此会在翻译时增加一些连词，如“同时”“而且”“所以”“因为”等。

Were there no gravity, there would be no air around the earth.

假如没有重力，地球周围就没有空气。(增译“假如”)

Since air has weight, it exerts force on any object immersed in it.

因为空气具有重量，所以处在空气中的任一物体都会受到空气的作用。(增译“所以”)

The emphasis on efficiency leads to the large, complex operations which are characteristics of engineering.

由于强调效率，因此引起工程所特有的工序烦琐而复杂的情况。

5. 增加表示抽象概念的词

如果句中的抽象名词有其具体的含义时，可根据上下文，补充适当的带有具体意义的名词，如“方法”“现象”“技术”“工程”“作用”“问题”“情况”等。

The lack of resistance in very cold metals may become useful in electronic computers.

这种在极低温金属中没有电阻的现象可能对电子计算机很有用处。(增译“现象”)

At low frequencies, the DC resistance of a given conductor is essentially the same as its AC resistance.

在低频情况下，给定导体的直流电阻实际上与其交流电阻一样。(增译“情况下”)

Resonance is often observed in nature.

在自然界中常常观察到共振现象。(增译“现象”)

6. 增加表示主语的词

在翻译时，遇到谓语“知道”“了解”“认为”等的动词时，可增加“我们”“人们”“有人”这样的代词，将被动句译成主动句。

It is recommended that a gauging with GO screw caliper gages should be supplemented by with the GO screw ring guage.

我们建议用通端螺纹卡测量时，应当用通端螺纹环规来补充。(增译“我们”)

With the development of modern electrical engineering, power can be transmitted to wherever it is needed.

随着现代化电气工程的发展，人们可以把电力输送到任何所需要的地方。(增译“人们”)

It has been thought that radium radiations might be useful in various diseases.

人们认为镭的射线可以用来治疗各种疾病。(增译“人们”)

Office Ergonomics

Since the first edition of this book have computers become ubiquitous, but also the very structure and definition of “office” has evolved. Desktop computers still basically have a keyboard, a screen, and a mouse; but each of these devices has morphed, often for the better, over the past 25 years. Keyboards and mice come in a variety of shapes and ***configurations*** and may be cordless. The screen has evolved into the flat panel that is more easily adjustable and much less prone to glare problems. Laptops are almost as ubiquitous as desktop computers, and handheld devices are commonplace. The concept of office as a place where everyone has a desk, a chair, and an assigned work area has often been replaced with a laptop that can be used anywhere, at home, in the car, or almost anywhere else.

Despite all these changes, ***Musculoskeletal Disorders*** (***MSDs***), also referred to as ***Cumulative*** Trauma Disorders (CTDs) or Repetitive Strain Injuries (RSIs), are unfortunately still with us. The main areas of the body involved with computer-associated injuries include the upper ***extremity***, the shoulder and neck, as well as the low back. In addition to MSDs, computer users often encounter eyestrain.

The variety of office equipment including chairs, keyboards, mice, and monitors has also expanded, and the Internet and networking groups have made access to this information available to almost everyone. Some of the products touted as ergonomic may not be well designed. Even well-designed products do not fit every situation. The interaction between computer and user is complex. Changing one variable affects all the other relationships. For example, changing the height of the seat pan on the chair affects how the upper extremity interacts with the keyboard and mouse, and how the head and eyes are positioned to view the display screen.

1. Chairs

The single most important piece of office furniture is arguably an easily adjustable chair. When people use computers for the majority of their time at work, it is essential that the chair is adjustable in height, provide ***lumbar*** support, and have an adjustable seat pan and backrest, as well as rounded edges to reduce contact stress. Armrests are desirable only if they are adequately adjustable. Chair adjustability is also important to avoid leg discomfort which is associated with ***static posture***, compression of the back of the thighs, and seat pans that are too high for the user.

Although many different chair designs are available, many do not meet even minimal standards. Helander et al. identify 10 separate features as essential in evaluating the functionality of an office chair, namely, seat height, seat depth, lumbar support height, lumbar support depth, armrest height, seat pan tilt, seat pan tilt tension, backrest height, backrest angle, and backrest tension.

A well-designed office chair minimizes leg discomfort associated with compression of the back of the thigh and postural immobility by providing rounded edges on the seat front as well as adequate cushioning over a firm support. Some research indicates that the seat height contributes to leg discomfort. For example, if the seat height is too high, it will compress the back of the thighs.

A chair designed to tilt, as opposed to a fixed position one, encourages continual small muscle changes to adjust posture. Some chairs hinge to follow the user through different positions; others are larger than necessary to allow the user to shift position often. Chairs that encourage continuous passive motion have been shown to reduce lower extremity swelling. The use of the stability ball in an office setting, however, was not recommended.

It is important to provide lumbar support in the sitting position. It is support restores at least some of the lumbar curve lost in the sitting position. Many chairs have a lumber support built into the seat back. For this support to be effective, it must be at the appropriate level, that is, at the level of the L5 disc. This is approximately at, or slightly below, belt level. A rolled towel or contoured pillow can be used in chairs without a lumbar support as an interim corrective measure. The thickness of the roll should be less than 2 in (1 in = 0.0254 m). Research indicates that back rolls that are too thick have resulted in increased muscle activity in the lower back. Chairs that provide adequate pelvic support also reduce muscle ***fatigue***.

One way to decrease the compressive force on the lumbar discs in the sitting position is to open the angle between the thighs and the torso. It is can be done by tilting the seat pan forward or reclining the backrest. Reclining the backrest at least 10° rotates the pelvis to help restore the lumbar curve and also transfers some of the torso weight to the back-rest. Harrison et al. (1999) report that seating with lumbar support and a backrest angle of 110°–130° provides the lowest compressive force on the lumbar discs as well as the lowest muscle activity in the ***spinal*** musculature.

Another way to transfer some of the body weight from the spine is to use armrests. These must be used at the proper height, which is approximately elbow height when the arms are hanging freely at the side of the torso. Chairs with armrests that are not adjustable may cause more problems than they solve. When armrests are too low, the person using the chair tends to slouch or bend to make the bodyfit the chair. If the armrests are too high, they may force the chair user to maintain abducted arms, which can contribute to neck and shoulder pain. The armrests also need to adjust low enough to clear the work surface so that the user can get positioned to use the keyboard in a neutral position.

All the best features in a chair will not contribute to worker health, comfort, and productivity if the chair does not fit the individual and is not adjusted properly. It is important to emphasize that not only should the chair be adjusted to each individual, but that each individual may need to make small adjustments periodically during the workday. The best designed chair works well only if workers understand the need to, as well as how to, adjust their chairs. In addition to being well designed, chairs should also be aesthetically appealing.

2. Keyboards

Traditional flat keyboards require close to full pronation of the forearm typically with ulnar ***deviation*** of both wrists for the fingers to reach the lateral keys. If the keyboard is too high, wrist extension may also contribute to ***awkward*** wrist posture. Ergonomically designed keyboards address these issues in a variety of ways. Some do this by splitting the keyboard in the middle; others are designed so that the keyboard is raised in the middle like a tent; still others may offer both configurations.

Because ulnar deviation of the right wrist to position the fingers over the number pad is one of the most awkward postures commonly seen, many keyboard designers have a detachable number pad. If the computer user needs to do a lot of work with numbers, the number pad can be repositioned so as to minimize ulnar deviation and other awkward postures of the upper extremity.

3. Mouse

Mouse use has increased with various computer applications including e-mail, CAD design, and web surfing. Many office workers use the mouse for half or more of the time they spend using a computer each day. The length of time using the mouse each day seems to have less impact than the overall amount of time spent using a computer.

Several different sizes and types of mice are available to use, but different mice are more useful for some applications than others. For example, CAD design requires precise placement control and some input devices do not provide that.

As we know reaching to hold the mouse positioned on a desk surface requires static awkward positioning of the upper extremity and the reporting of increased musculoskeletal symptoms. The awkward postures include abduction of the arm, activation of the forearm/hand extensor muscles, full pronation of the forearm, and, quite often, ulnar deviation. These positions can also contribute to musculoskeletal symptoms in the shoulder and neck. When the right hand is used to operate the mouse, adverse musculoskeletal effects can be exacerbated because the number pad to the right of the alphabet section of the keyboard requires the mouse be placed even further to the right requiring increasingly awkward posture.

Positioning the mouse lower and toward the midline of the user can help to reduce postural stress. Some keyboards do this by attaching a mouse pad that can be positioned along the edge facing the user. Use of a touch pad or roller device that can be positioned in the center between the computer user and the keyboard also achieves this goal. Forearm support while mousing is another way to reduce musculoskeletal stress.

Rotation of the typical two button mouse so that it is oriented vertical rather than flat with respect to the desk is another way to reduce full pronation and ulnar deviation.

4. Monitors

Most humans prefer to read from paper than from display screen for many reasons. Resolution, measured in dots per inch (dpi), is one factor that affects readability. Paper provides a resolution of about 225 dpi. Display screens typically provide resolutions between 60 and 120 dpi. High-resolution monitors may promote better performance, but they also produce smaller images that may be perceived as harder to read.

Flicker is dependent on the refresh rate of the display screen and is measured in cycles per second or hertz (Hz). Humans can consciously detect flicker at rates below 50-60 Hz. Flicker in the 10 Hz frequency may precipitate seizures in epileptics. Head-aches and blurred vision have been associated with flicker rates that are not consciously perceived, but nonetheless have a physiological effect.

Computer users need to know what adjustments they can make on their display screens to adjust focus, brightness, and contrast. Focus is affected by internal and external adjustments. If the image seems blurred, screen focus should be checked first. It may be caused by too high settings of brightness or contrast, or it might be the result of internal problems. Font size and refresh rate can be adjusted through internal computer settings. Special so ware programs are available that can be used by individuals with vision impairments.

Flat-panel display units are replacing the cathode ray tube monitors that were in use in the 1990s. The newer monitors control glare very well and are easier to position for both height and slant.

5. Desks

Typical office desks are 29-30 in. high. Although this works well for many tasks, such as reading, writing on paper, and talking on the phone, it is not an appropriate height for keying or mousing for most people. Unless the workstation is to be used exclusively for computer work, the traditional surface height works well for most people. The exceptions are very tall or very short people, the two ends of the bell-shaped curve for height distribution.

Many types of modular furniture have adjustable desk surface heights, but the adjustment requires tools and effort and is, therefore, not an option for small or daily changes. It does, however, allow for a change in height of the work surface when the occupant of the workstation is reassigned.

Another option for changing the work surface height is an easily adjustable worktable. The adjustment can be done either with a crank or a hydrolic mechanism. The entire work surface with all its contents, keyboard, mouse, monitor, phone, paper, and whatever else is on the desk, can be adjusted up or down during the course of the workday. Another advantage of the adjustable height table is that it allows the computer user to easily change from working in a seated position to working in a standing position. The adjustable height worktables are expensive.

6. Keyboard Trays

There are a variety of keyboard and mouse trays available. The first step is to make sure that the mechanism that attaches the keyboard tray to the underside of the desk is easily adjustable for both height and slant, and sturdy enough to hold the position set by the computer user. Then users have to become comfortable with their ability to make small adjustments to the keyboard tray throughout the work day.

New Words and Expressions

configuration [kənˌfɪɡəˈreɪʃ(ə)n]	*n.* 配置，结构，外形，组合
Musculoskeletal Disorders (MSDs)	肌肉骨骼疾病
cumulative [ˈkjuːmjʊlətɪv]	*adj.* （在力量或重要性方面）聚积的，累计的，累积的
extremity [ɪkˈstremɪti]	*n.* 末端，极端，极限，极度
lumbar [ˈlʌmbə]	*n.* 腰椎，[解] 腰动脉，腰神经
static posture	静态姿势，静止姿势，静止体位
fatigue [fəˈtiːɡ]	*n.* 疲劳，劳累，厌倦，（金属或木材的）疲劳
spinal [ˈspaɪn(ə)l]	*adj.* 脊的，脊椎的，脊髓的
deviation [diːvɪˈeɪʃ(ə)n]	*n.* 偏差，偏离，背离，违背
awkward [ˈɔːkwəd]	*adj.* 令人尴尬的，使人难堪的，难对付的，难处理的

Unit Five

5

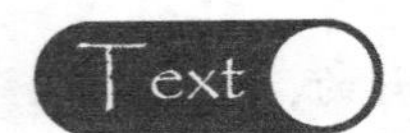

An Overview on Safety Law

While you are at work, in whatever location or environment that may be (e. g. on a building site or in a client's home), you need to be aware of that there are safety laws to protect you from harm. The laws state how you should be protected and what your employer has to do to keep you safe, i. e. their responsibilities. Health and safety legislation not only protects you, but also states what your responsibilities are in order to keep others safe. It is very important that you follow any guidance given to you regarding health and safety and that you know what your responsibilities are.

1. The Legal Framework for Health and Safety

There are two sub-divisions of the law that apply to health and safety issues: criminal law and civil law. Criminal law consists of rules of behavior laid down by the Government or the State and, normally, enacted by Parliament through Acts of Parliament. These rules or Acts are imposed on the people for the protection of the people. Criminal law is enforced by several different Government Agencies who may ***prosecute*** individuals and organizations for contravening criminal laws. One example of criminal law enforcement agency is The Health and Safety at Work (HSW). Act is an example of criminal law and this is enforced either by the Health and Safety Executive (HSE) or Local Authority Environmental Health Officers (EHOs). Other agencies which enforce criminal law include the Fire Authority, the Environment Agency, Trading Standards and Customs and Excise.

Civil law concerns disputes between individuals or individuals and companies. An individual sues another individual or company to address a civil wrong or tort.

2. Legislation of Occupational Health and Safety

Safety laws vary with country especially at home and abroad. But occupational health and safety legislation in most of countries has a similar framework—the Robens model, one kind of self-regulation style legislation. Numbers of states either adopt the Robens model completely or in part.

In the UK, the original occupational health and safety legislation was based on *the English Factory Acts* (1833 – 1894) and *the Workers' Compensation Act* of 1897. These factory acts were developed to help improve the poor working conditions in factories and mines that had developed during the Industrial Revolution. *The English Factory Act* style of legislation is ***prescriptive*** and requires the attainment of minimum standards and the periodic inspection of the workplace. In 1972 a new style of legislation was introduced in South Australia, and in 1974 in the United Kingdom, as the result of the findings of the Robens Committee—*The Health and Safety at Work Act*. The Robens Committee criticized a system of too much law administered by too many authorities based on the enforcement of minimum standards by an inspectorate. *The Health and Safety at Work Act* provides for self-regulation by industry and does so by involving those most at risk in the workplace—the workers. An emphasis was placed on prevention, rather than inflicting heavy penalties as punishment, although serious penalties were retained for gross breaches of the law. *The Health and Safety at Work Act* also consolidated many scattered pieces of legislation, enabling occupational health and safety to be focused into one act. Standards and regulatory requirements must still be met, but with the opportunity for worker participation and involvement.

The United States didn't completely adopt the Robens model. Until 1970 occupational health and safety requirements were enforced at a state level. The US Congress, passed *the Williams-Steiger Occupational Safety and Health Act* (*OSH Act*). A new administration was set up, the Occupational Safety and Health Administration (OSHA), part of the Department of Labor. This works in conjunction with the US Department of Health and Human Services which is responsible for the National Institute of Occupational Safety and Health (NIOSH), part of the Centers for Disease Control. The act allowed national collection of Occupational Safety and Health (OSH) statistics for the first time. Occupational safety and health standards are put out by the Department of Labor and take legal precedence over state laws and regulations. However, states can continue to administer OSH legislation as long as it meets federal standards. Unlike the British legislation, the US legislation has adopted, until now, a prescriptive approach, with a strong emphasis on compliance and citations for violations. The concept of health and safety representatives is not in the legislation. In 1982 a form of self-regulation was introduced. The Voluntary Protection Program allowed best practice employers reduced ***penalties*** and exemption from inspection. And, MSHA, also part of the Department of Labor, enforces OSH in the US mines under *the Mine Safety and Health Act*, but again with research and other support from NIOSH, such as miners' medical surveillance.

3. Features of the Robens Model

Tripartite policy-making body: As part of the Robens approach, policy making in OSH has passed from a traditional public service department to a body with representation from government, the unions, employers, and, in some cases, independent experts. In the Canadian Province of Alberta, for example, S. 6 of the *OSH Act* sets up an OSH Council. The USA has not generally adopted the Robens principles, but does have under S. 7 of the *OSH Act* a multipartite National Advisory Committee on OSH. Unified administration: The Robens recommendation for a unified administration has not been accepted in all those jurisdictions. Specialist areas such as the nuclear industry may operate under the prime OSH legislation but be regulated by a specialist inspectorate under an agency agreement with the prime OSH authority. But the US, not totally Robens based, has two principal OSH administrations—the Occupational Safety and Health Administration, OSHA and the ***Mine Safety and Health Administration***, MSHA.

Coverage of all workplaces: If all OSH legislation is considered, nearly all workers are now covered by one piece of OSH legislation or another. Special provisos may apply to coverage of military personnel. For example, under *the Australian OSH (Commonwealth Employment) Act*, the Chief of the Defence Forces can declare exemptions. Most jurisdictions, however, now have no problem including police within the coverage of the OSH legislation.

Self-regulation: Self-regulation has been chosen as the main strategy in improving OSH standards in the workplace. Self-regulation moved us away from the inspectorate-driven, prescriptive legislation of the 20th century and three quarters, to a new model based primarily around a general duty of care by various stakeholders in the workplace, and to consultation, cooperation, information and training. Fines and penalties for breach of duty under Robens legislation remain.

General duty of care: A key concept in Robens-style legislation is to take the concept of the general duty of care developed through common law tort cases (and already in some factories legislation) and apply it to a variety of stakeholders in the workplace.

Because of the form of legislation, safety laws, in a broad sense, contains kinds of different forms of regulations. A fuller description follows.

Codes of practice: Codes of practice are used in some countries. They are not legislation, but they are developed to provide practical assistance to people to comply with the requirements of OSH and other acts and regulations on a workable basis.

National government standards: These are documents issued by an OSH authority which set standards in particular areas, e. g. hazardous chemicals. Their legal effect varies between jurisdictions. They should not be confused with standards issued by standards organizations.

Standards issued by standards organizations: There are a large number of national standards relevant to OSH issued by organizations such as the ANSI, BSI, Canadian Standards Association, Standards Australia, Standards New Zealand and so on, sometimes referred to as ***consensus*** standards. They may be given some legal status by the OSH laws or regulations—if so, they are said to be "picked up" by the legislation. International standards may also be referred to, e. g. ISO, CIE, IEC and ICRP.

Guidance notes: Some OSH authorities may issue a guidance note or guideline which is an explanatory document providing detailed information on the requirements of legislation, regulations, standards, codes of practice or matters relating to occupational safety and health.

Industrial agreements: Industrial or collective bargaining agreements, or in Australia awards made by an industrial commission, detail the conditions of work and often include occupational safety and health provisions. There may also be agreements on issues such as the resolution of disputes over dangerous work and the right to stop work with pay, although this issue is also covered in some OSH legislation.

Occupational safety and health were previously regulated separately by two government departments. As there are some overlaps at the one hand, and omission at the other, regulatory functions on occupational safety and health started to see some emerging trend recently since 2012, when the Safety Production Supervisory Commission is entitled to regulate and manage both safe and occupational health issues nationwide.

Since the 1990s, there is continuing development of safety related legislation. The foundation law of safety legislation is *the Labor Law* **promulgated** July 5, 1994, and was effective as of January 1, 1995. The Law is the basic body for adjudication of labor relations, and has established labor contract and group contract systems, a tripartite coordination mechanism, a labor standard system, a system for handling disputes, and a labor supervisory system, basically shaping a new approach to labor relations. Since the 1994 *Labor Law*, several other important laws and regulations have come into force (see Table 5-1).

Table 5-1 A Brief Guide to Some OSH Legislation and Organizations

Countries	Legislation (Not Including Workers' Compensation)	Administration	Professional Organizations	Professional Qualifying Organizations	Other Organizations
The US	*Occupational Safety and Health Act*, *Mine Safety and Health Act*, and stander gazette in Consolidated Federal Register	Nursing Society of Occupational Health Nursing, Department of Labor-Occupational Safety and Health Administration and Mine Safe and Health Administration, Environmental Protection Agency, and in some states, state OSH Departments, National research and standards recommendations: National Institute of Occupational Safety and Health, and Bureau of Mines	American Society of Safety Engineers, American College of Occupational and Environmental Medicine, American Conference of Governmental Industrial Hygienists, American Industrial Hygiene Association, Human Factors and Ergonomics Society, American Association of Occupational National Institute of Health Nurses, Acoustical Society of America	Board of Certified Safety Professionals, American Board of Industrial Hygiene, American Board for Occupational Health Nursing	
The UK	*Health and Safety at Work Act*, *Nuclear Installations Act*, and the others, and Regulations	Health and Safety Institute Executive assisted by specialist agencies (e. g. nuclear) and Health and Safety Commission	Institute of Occupational Safety and Health, Institute of Occupational Medicine, British Occupational Hygiene Society, The Ergonomics Society,	National Board Safety of Occupational Health, British Examining and Registration, Board in Occupational Hygiene, Faculty of Occupational Medicine of the Royal College of Physicians	Royal Society for of Physicians the Prevention of Accidents

New Words and Expressions

prosecute [ˈprɒsɪkjuːt] v. 起诉，控告，检举，担任控方律师
prescriptive [prɪˈskrɪptɪv] adj. 指定的，规定的，规范的，约定俗成的
penalty [ˈpen（ə）lti] n. 惩罚，罚款，（犯规）处罚，罚点
Mine Safety and Health Administration 矿山安全和健康管理局
coverage [ˈkʌv（ə）rɪdʒ] n. 新闻报道，信息质量，提供的数量，覆盖范围（或方式）
codes of practice 行为守则，业物法规
consensus [kənˈsensəs] n. 一致的意见，共识
promulgate [ˈprɒm（ə）lgeɪt] v. 颁布，传播，宣传，发表

Exercises

1. What is safety law? How can safety law protect you from harm?
2. What are the two sub-divisions of the law that apply to health and safety issues? Please explain how they work and the relationship between them.
3. Explain the occupational health and safety legislation.
4. What is Robens model? What are the features of the Robens model?
5. What is the main strategy in improving occupational health and safety standards in the workplace?

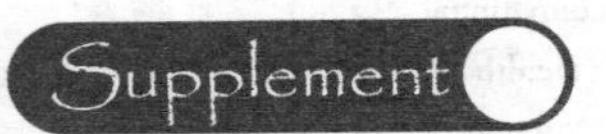

科技英语翻译技巧——词语的省略

英语和汉语的用词特点不同，在英语中常用的词有时在汉语中是可以省略的，因此，在翻译时，可省略在汉语中多余的词语，以便使译文更加精炼、准确。

1. 省略冠词

在英语的用词习惯中，很多冠词的使用是没有必要的，在英译汉时可以省略。在常用的英语词组中，使用冠词可以改变词组的意义，翻译时，可不译出冠词本身的意思，只注意冠词引起词组的变化即可。在以下情况可以省略冠词：

1）涉及一般规律的不定冠词。

2）泛指类别的定冠词。

3）用于限定词语的定冠词。

4）用于形容词最高级的定冠词。

The resistance of a conductor is closely related with its length, its cross-sectional area, and *the* material of which.

导体的电阻与其长度、截面积和制造材料密切相关。(省略冠词“the”)

The first thing that a factory should do is to increase the quality of products.

工厂的首要任务是提高产品质量。(省略冠词“the”)

还有冠词不能省略的情况：

1）表述数量的不定冠词。

2）表示“每一”“同一”的不定冠词。

3）特指或重复前面事物的定冠词。

When *a* free electron meets such *a* hole in its motion, it is possible for *the* electron to occupy the hole and thus a pair of electron and hole disappears.

当一个自由电子在运动中碰到这样一个空穴时，该电子可能占据这个空穴，而使一个电子空穴对消失。

2. 省略代词

在汉语里，若句子前后同为一主语时，可省略后面的主语。因此，在翻译时常常可省略原句中重复出现的代词。

（1）省略人称代词

Electrical leakage will cause a fire, hence *you* must take good care of *it*.

漏电会引起火灾，因此必须好好注意。（省译“you”“it”）

（2）省略物主代词

According to *their* ability to conduct electric current, all materials may be classified into three major categories: conductors, semiconductors, and insulators.

根据导电性能的不同，所有材料均可分为三大类：导体、半导体和绝缘体。（省略“their”）

（3）省略反身代词

A gas distributed *itself* uniformly through a container.

气体均匀地分布在整个容器中。（省略“itself”）

（4）省略指示代词

The volume of the sun is about 1, 300,000 times *that* of the earth.

太阳的体积约为地球的130 万倍。（省略“that”）

3. 省略引导词

（1）可省略做形式主语、强调语气、指示代词的 it

It is the gravitation which makes the satellites move round the earth.

地球引力使卫星绕地球运行。（省略“it”）

（2）省略 there

There exist neither perfect insulators nor perfect conductors.

既没有理想的绝缘体，也没有理想的导体。（省略“there”）

4. 省略动词

英语中的谓语都是动词，但是在汉语里，谓语不仅可以是动词，还可以是名词、形容词、介词结构、主谓结构等。因此，在翻译时，原句中的谓语动词可以省略。

In the case of a possible accident, the machine *comes* to a halt immediately upon the operator's direction.

一旦发生事故，只要操作人员一发指令，机器就立刻停下来。（省略“comes”）

Hence, television signals *have* a short range.

因此，电信号的传送距离很短。（省略“have”）

Now the steam formed *has* the same pressure as the water has.

现在所形成的蒸汽压力与水的压力相等。（省略“has”）

5. 省略介词

英语中常常使用介词，介词往往是联系词与词的桥梁。汉语中词与词之间是通过语序和逻辑关系来表示的，在翻译时常常可以省略原句中的介词。

When a substance changes only *in* state or *in* form, it is a physical change.

当一种物质只改变状态或形式时，就是物理变化。(省略“in”)

The temperature of the liquid is raised *by* the application of heat.

加热可以提高液体温度。(省略“by”)

The ionization in the upper atmosphere is caused by ultra-violet rays *from* the sun.

上层大气中的电离作用是由太阳紫外线引起的。(省略“from”)

6. 省略连词

英语中的连词是用来连接句子或词语的逻辑关系的词，起着承上启下的作用，在翻译时，往往可以省略。

The system is totally enclosed *and* prevents air pollution from dust particles or gases.

此系统为全封闭系统，能防止尘粒或烟气污染大气。(省略“and”)

The average speed of all molecules remains the same *as long as* the temperatures is constant.

温度不变，所有分子的平均速度也就不变。(省略“as long as”)

The transistor gate has a distinct advantage over the diode gate *in that* the transistor amplifies, as well as acting as a gate.

晶体管门电路比起二极管门电路来有一个显著的优点，那就是晶体管除了起门的作用之外还能够放大。(省略“that”)

Developments of Safety Law

In industrialized countries, many businesses are replacing direct employment and turning to subcontractors, labor hire agencies and supply chains to perform work previously done by employees of the business. Laws originally intended to protect workers from health and safety risks are not designed to address these situations. And these laws are not well adapted to accommodate the changing world of work that has left the status of many workers unclear. The employment contract, which was central in defining the scope of the general duty clause under Australian Occupational Health and Safety (OHS) laws enacted in the mid-20th century, is no longer a relevant concept. There is now a ***substantial*** body of evidence showing that precarious work relationships are associated with more hazardous working conditions. In order to adapt labor laws to these emergent work arrangements, different authors have suggested considering the objectives of legislation and focusing on the aspects of employee rights and well-being that it aims to protect. According to this perspective, OHS laws should provide broad-spectrum protection, covering all forms of employment relationships, in order to protect the fundamental human right to health and prevent of work-related injury and disease. OHS laws have to be adapted to the new forms of work relationships in order to meet their primary objective, namely, the protection of the human right to health.

1. A New Volunteers and the *Health and Safety at Work Act* in New Zealand

The ***Health and Safety Reform Bill*** has passed to create a new *Health and Safety at Work Act*, which will come into force on 4 April, 2016. Under the Act, the coverage of volunteers reflects what it is under the current law which distinguishes between casual volunteers and volunteer workers. This recognizes that volunteers contribute greatly to New Zealand communities and will ensure the new law will not negatively affect volunteering. The information below explains how the Act applies to volunteers.

2. Is your organization a Person Conducting a Business or Undertaking?

Under the Act, a Person Conducting a Business or Undertaking (PCBU) has the primary duty to ensure the health and safety of its workers and others, so far as is reasonably practicable. A purely volunteer organization where volunteers work together for community purposes and which does not have any employees is known as a volunteer association under the Act. A volunteer association is not a PCBU so the Act will not apply to it. A volunteer organization which has one or more employees is a PCBU and will have the same duties as a PCBU to ensure, so far as reasonably practicable, the health and safety of its workers and others. There are some exclusions to this, depending on whether the PCBU has ***casual*** volunteers or volunteer workers. This is the same approach as taken by the current law. What the volunteer organization will have to do is what is reasonably practicable for it to do, and what is within its influence and control.

3. If your organization is a PCBU, does it have casual volunteers or volunteer workers?

Where volunteers carry out work for a PCBU, the Act distinguishes between casual volunteers and volunteer workers. Volunteer workers are people who regularly work for a PCBU with its knowledge and consent on an ongoing basis and are integral to the PCBU's operations. This ***distinction*** is based on the existing *Health and Safety in Employment Act* 1992. PCBU's will owe a duty to ensure, so far as reasonably practicable, the health and safety of volunteer workers (as if they were any other worker). This ensures that these volunteers are afforded the protection of having the appropriate training, instruction or supervision needed to undertake their work safety—just like any other worker.

4. What are a PCBU's duties to others?

PCBU's will have a duty to others (such as customers or visitors) to ensure that their health and safety is not put at risk from the PCBU's work, so far as is reasonably practicable. This duty also applies to casual volunteers.

5. Are your volunteers doing certain activities which means they are excluded from the "volunteer worker" definition under the new law?

People volunteering for the following activities will not be volunteer workers under the new law:

- Participation in a fundraising activity assistance with sports or recreation for an educational institute, sports or recreation club.

- Assistance with activities for an educational institution outside the premises of the educational institution.

- Providing care for another person in the volunteer's home. This approach follows the existing *Health and Safety in Employment Act* 1992. Although casual volunteers and volunteers doing these activities won't be "volunteer workers" for the purposes of the Act, their health and safety will still be covered by the PCBU's duty to other persons affected by the work of the business or undertaking.

6. Changing Work Relationships and the Protection of Workers under Australian Occupational Health and Safety Law

Australia, like Canada, is a federation consisting of three tiers of government: a federal government, state and territory governments, and local governments. In 2008, the Council of Australian Governments formally committed to harmonising OHS regulation through an Inter-Governmental Agreement. An extensive National Review into the structure and content of model OHS laws was conducted, and was required to take into account the changing nature of work organization and work relationships. A model *Work Health and Safety Bill* (the Model Act) was finally adopted, with each jurisdiction agreeing to enact "mirror" legislation and codes of practice.

One of the Model Act's most important reforms was to reframe the general duty clause in a way that departed from the paradigm of the traditional employment relationship. To date, this reform has been accepted by all jurisdictions in their harmonised *Work Health and Safety Acts*. Under the harmonised legislation, the "primary duty of care" lies with a PCBU, who must ensure, ***so far as*** is reasonably practicable, the health and safety of all types of workers "engaged", "caused to be engaged", "influenced" or "directed" by the PCBU. The concept of PCBU is defined very broadly, as highlighted in the guidelines, and includes both individuals and organizations. The phrase "business or undertaking" is intended to be expansive, covering businesses or undertakings conducted by persons including employers, principal contractors, head contractors, franchisors and the Crown. A "worker" is described as a "person who carries out work in any capacity for a person conducting a business or undertaking". This extensive definition of worker, as opposed to employee, captures the wide ***array*** of complex work relationships.

In 2012, the Industrial Relations Commission of New South Wales considered the meaning of the term "worker" under the *Work Health and Safety Act* 2011 (NSW) for the first time. The Commission accepted that a PCBU's OHS obligations extended to employees of subcontractors, even though there was no direct contractual relationship between the PCBU and the subcontractor. In order to fulfil the duty of care owed to workers, a PCBU must ensure, so far as is reasonably practicable, the following: the provision and maintenance of a safe environment, plant and systems of work; the safe use, handling, storage and transport of substances; the provision of adequate facilities for the welfare of workers; the provision of proper information, training, instruction or supervision; and the monitoring of the health of workers and conditions at the workplace.

Another important reform is the new positive duty imposed on an officer. An officer of a PCBU subject to a duty or obligation under the Act must exercise due diligence to ensure that the PCBU complies with that duty or obligation.

As well, the definition of "due ***diligence***" is highly inclusive and largely consistent with the duty of care and diligence owed by directors and officers under *the Commonwealth Corporations Act* 2001. It requires officers to, amongst other things, acquire and keep up-to-date knowledge of OHS matters, generally understand the health and safety risks emanating from the PCBU's operations, ***warrant*** that proper resources are available to eliminate or minimise those risks, and ensure that proper health and safety management systems are in place. This duty is personal and cannot be delegated. Consequently, officers can be convicted for failure to exercise due diligence, independently of whether the corporation has or has not complied with its own duties. Notably, they cannot discharge their duty by relying on the ***compliance*** of another person.

The Australian model offers a different solution, one that recasts the duty of care so that it applies to all persons engaged in the conduct of a business or undertaking. When multiple PCBU's are involved, they have an obligation to consult, cooperate and ***coordinate*** in order to ensure the health and safety of all categories of workers. Moreover, the broad definition of "worker" recognizes the changing nature of work relationships and ensures OHS protection regardless of the existence of an employment contract. In fact, the use of the term "work" rather than "employment" throughout *the Work Health and Safety Acts* illustrates a paradigm shift designed to cover a wide array of complex work relationships. These workers also have the right to participate in workplace arrangements for OHS.

New Words and Expressions

substantial [səbˈstænʃ(ə)l] *adj.* 大量的，结实的，牢固的，重大的

Health and Safety Reform Bill 健康与安全改革法案

casual [ˈkæʒjʊəl] *n.* 临时工，便装，便鞋

distinction [dɪˈstɪŋ(k)ʃ(ə)n] *n.* 区别，区分，差别，卓越

so far as 只要，就……而言，在……范围内

array [əˈreɪ] *n.* 数组，阵列，大量，大堆

diligence [ˈdɪlɪdʒ(ə)ns] *n.* 勤奋，勤勉，用功

warrant [ˈwɒr(ə)nt] *n.* 依据，许可证，执行令，授权令

compliance [kəmˈplaɪəns] *n.* 服从，听从，承诺，柔软度

coordinate [kəʊˈɔːdɪneɪt] *n.* 坐标，（颜色协调的）配套服装

Unit Six

Safety Training and Teaching

A barge was undergoing repair and maintenance in Singapore. A worker entering a ballast tank lost consciousness. Two other workers who went in to rescue him were similarly affected.

All three were rescued and recovered. The tank had been closed before the accident and was rusted inside. Rusting consumes oxygen. The oxygen level in an ***adjacent*** unopened tank was tested and the result was 4.5%. The normal level of oxygen in air is 21%. The barge owner was fined as there were no safe work procedures and permit-to-work system as the law required. A second fine was applied because the workers had not been sent to a safety training course which had the Chief Inspector of Factories' approval.

1. Education and Training Needs

Employee education aims to develop knowledge and understanding, rather than knowledge and skill, for a defined activity through various methods which provide an understanding of traditions, ideas and concepts. It involves verbal ***as well as*** other communication channels which are fundamental to learning. Education on matters such as safety, hazard management, and emergency procedures is vital to safety management.

Training, on the other hand, is the planned and systematic sequence of instruction, under competent supervision, designed to develop or improve the predetermined skills, knowledge and abilities required by an individual to perform a task to a particular standard. Training can involve various techniques including on the job coaching or mentoring, demonstrations, group or individual exercises, role playing, case studies and displays to name a few. Some can be on-the-job. An example of on-the-job training would be in accident investigation.

2. Role of Health and Safety Training in Safety Management

Effective health and safety training supports organizational objectives and plays an important role in

safety management. The management system creates a safety culture enhancing occupational safety and health that reinforces safe and healthy work practices while training helps provide the knowledge, skills and practice necessary to sustain this culture. The objectives of health and safety management cannot be achieved without systematic training to identify, assess and control hazards and to put safety in the forefront of every person, activity and situation by using defined methods. Training creates an environment within which positive change can occur and provides a ***forum*** for discussion to improve performance.

3. Current Approaches to Training

There is an important emphasis now in a number of countries on competency-based training, with an emphasis on getting away from the training room and assessing people's actual performance on the job. There may be special units in OHS written for nationally recognized training packages for different industries and for different levels of supervision and management. The Australian National Occupational Health and Safety Commission, for example, has also developed generic competencies in OHS for employees, supervisors and managers, which are a useful basis for integrating OHS into other training, and are now incorporated in some national industry training packages.

(1) Inductions

A training-needs analysis allows ***optimal*** use of money and resources. A lot of training plans follow from job descriptions (including future roles for employees) and from procedures.

Induction training for all new employees is especially important. Training needs to consider cultural differences in perception of risk, and the needs of workers who don't speak the local commonly used language. In industries with a high turnover such as mining or construction, a common OHS induction ***module*** has been developed in some countries, e.g. Australia, leading to a ***portable*** qualification recognized across the industry. The UK is considering such a "passport". In Australia the module is based on a core OHS module which has been developed to sit within "training packages" for different industries.

(2) Writing Lesson Plans

Lesson plans outline what is to be taught and the methods to be used in a standardized format. They can be used in the traditional training approach, but also in on-the-job methods. They are written to help the instructor to:

- present material in the proper order;
- avoid omission of essential material;
- conduct the sessions according to a timetable;
- place proper emphasis on items to be covered;
- provide for trainee participation;
- gain confidence;
- assemble all equipment needed.

Lesson plans should be written so that information can be viewed easily; that is, handwrite or type legibly allowing enough space to insert breaks, messages, overheads, etc., during the course of training. It is important to remember that lesson plans are not a script; they should contain only the bare minimum of what is required; pertinent information to be revealed; and what is to be accomplished at the end of each segment. A variety of lesson plan formats exist. It is best to ***draw up*** a lesson plan that you are comfortable with by integrating the best features of different lesson plans that you have at your disposal.

The bottom line when designing lesson plans is to:

- keep information flowing in an ordered fashion;
- have a clear objective of the message you want to convey to the trainees by the end of the session;
- keep the participants interested.

(3) Preparing a Training Register

It is essential that a register recording employee participation in training programme is kept so that new or untrained workers are identified; those needing retraining are listed; and proof exists that the company has fulfilled its obligation under training and occupational health and safety legislation to provide adequate information, instruction and training to workers. As a minimum, the record should contain the trainee's name, the job for which he/she is being trained, the date and period of time when training took place, the training given, and comments by the trainer or instructor. Advantages of this approach are its usefulness when considering the future development of the individual and possible improvements to the training system.

4. Training Needs Analysis, and Design, Conduct and Evaluation of an OHS Training Programme

(1) Undertaking a Training-Needs Analysis

Before going further into this matter, we need to focus on the words "for a given workplace". The reason is that we will assume that training-needs analysis at the organizational level has been carried out. This first requires that areas of priority for the organization have been analysed to see if they are necessary, and then job descriptions have been prepared. Another way of looking at it is to consider that a number of tasks are required to be undertaken by the organization to achieve its objectives. Once we have decided what the tasks are and how we divide them up into manageable jobs for people, we can write a job description. This is an important step, because multiskilling requires us to look at which tasks will be assigned to a job rather than asking, for example for a carpenter's job, what the tasks are. Sometimes, preparing the job description allows us to decide if a particular occupation is necessary for the organization, and this may feed back into re-evaluating the way tasks are divided up.

If the process above has taken place, and the job descriptions have been written, then the knowledge and skills needed to carry out particular tasks can be identified in the job specifications. Skills may be cognitive, affective or psychomotor skills—that is, reasoning, sensitivity or feeling (as for a musician) or effective movements, respectively. You will need to identify ***aptitudes*** and particular personal traits which suit the person to the job.

If you have a wide range of people to choose from, the job specification need not contain too many fundamental items of knowledge and skill. This is the role of personnel selection, e. g. if you don't want to have to train people in a certain level of reading skills, you could select accordingly. If you want to select people with fundamentally good attitudes to safety, good targeted selection questions can help you to do that. On the other hand, aiming to provide job opportunities for disadvantaged groups may require attention to fundamentals, such as literacy.

The next step is to compare the knowledge and skills the job specification has with the existing knowledge and skills of the person who will do the job. This is called gap analysis. The training needs are then identified and training objectives set.

You can prepare questionnaires or forms to analyse trainees' characteristics and to analyse a trainee's current knowledge and skills against the different job requirements. This will tell you the training required. You can then write training objectives. Job procedures are a good basis for designing training.

(2) Designing an OHS Training Programme

The programme needs to meet the organization's individual needs, and apply the legal requirements for health and safety training to the programme's design.

1) Objectives. The training objectives should describe the type of behavioural change which will occur; that is, the skills which will be observed on the job. The objectives show:

- what has to be done;
- under what conditions, and with what;
- to what standards.

And must be accompanied by an effective way of evaluating the results of the training. This approach is known as Competency-Based Training (CBT), mentioned earlier.

2) Induction training. Delivery and costs and their interrelationship were mentioned above. For induction training, for example, people may be coming on site daily. It is no use running induction training once a fortnight, so the organization may be looking for a self-paced package with built-in assessment so that people can do it on any day at any time. In the interactive multimedia mode of delivery this can be fairly expensive, but may still be a cost-effective option for some organizations. Simpler print-based computer packages and print and paper-based materials may be sufficient. Some industries now have generic inductions with portability (that is, mutual recognition), and then only an add on site and workplace specific induction may be required.

The same content can be delivered on-and off-the-job. On-the-job may involve one on-one training under a supervisor or other experienced employee, but these should have had "train-the-trainer" training, i. e. training in how to train, and assess competency.

3) Legal aspects. The legal aspects are going to vary depending on the type and level of training and the type of trainee. Training for an open-cut mine manager, for example, would require extensive knowledge of the mining safety legislation.

For anyone in a workplace the minimum requirements in the training should include:

- appropriate knowledge of legislation;
- knowledge of employee entitlements under the legislation—training, including safe procedures, safe systems, safe workplace (so far as is reasonably practicable), election of safety and health representatives where applicable, consultation, right to refuse unsafe work, and information, e. g. MSDS;
- duties of employees;
- reporting accidents;
- hazard identification and risk assessment, e. g. job safety analysis;
- specific hazards relevant to the workplace concerned.

4) The training course itself. It is necessary to emphasize here once again that a training course need not consist entirely of trainees sitting and doing activities in a training room. This might be all of, part of, or none of the course. There might be a face-to-face off-the-job ***segment***, an on-the-job segment and some individual work at a time selected by the trainee, all of which is part of the course.

(3) Conducting a Training Session

Within a training course conducted in a training room, you need to vary the training techniques to suit the objectives, which may include knowledge transfer; problem solving; skills development (of which problem solving is a part); and change in attitudes. You may decide to move out of the training room to do a hazards identification exercise in the workplace, remembering the safety of your trainees. Some training techniques are set out below.

To transfer knowledge, use:

- group discussions (questions and answers);
- group or individual exercises;
- lectures (with ***handouts***);
- forums;
- panel discussions;
- films, videos, etc.

To practise problem solving, use:

- case studies;
- brainstorming;
- discussion groups;
- exercises, etc.

To develop skills, use:

- demonstrations for manual skills;
- role playing for interpersonal skills;
- peer teaching;
- programmed instructions, etc.

To change attitudes, use:

- debates;
- displays;
- role playing (for clarifying how others feel);
- group discussion (for group attitudes);
- individual exercises;
- demonstrations;
- campaigns, etc.

Behaviour, including safety behaviour, is influenced by attitudes, but often it is easier to change behaviour than attitudes.

You should now be able to put together a lesson plan, overheads or computer slide show and handouts for the training room segment of your training programme and present the segment.

Frank Bird Jr. suggests you keep five Ps in mind in a presentation: prepare, personalize, picturize, pinpoint, and prescribe. He also suggested FIDO—frequency—how often; intensity—how vivid; duration—how long; and over-and-over—spaced repetition of the issue.

In countries such as Australia you will also find assistance in materials which have been produced under the national training agenda. These include OHS competencies for specific occupational areas and levels of responsibility. In the UK, the New Vocational Qualifications relating to safety are of assistance.

New Words and Expressions

adjacent [əˈdʒeɪs（ə）nt]	*adj.* 邻近的，毗连的
as well as	也，和……一样，不但……而且
safety management	安全管理
forum [ˈfɔːrəm]	*n.* 论坛，讨论会，法庭，公开讨论的广场
optimal [ˈɒptɪm（ə）l]	*adj.* 最佳的，最理想的
module [ˈmɒdjuːl]	*n.* ［计］模块，组件，模数
portable [ˈpɔːtəb（ə）l]	*adj.* 手提的，便携式的，轻便的
draw up	草拟，起草，停住，使靠近
aptitude [ˈæptɪtjuːd]	*n.* （学习方面的）才能，资质，天资
segment [ˈsegm（ə）nt]	*n.* 段，部分，片，弓形
handout [ˈhændˌaʊt]	*n.* 讲义，施舍物，救济品

Exercises

1. What is the significance of training and teaching?
2. What does the employee education aim to? Why do employees need education?
3. What is the role of health and safety training in safety management?
4. How many steps does an OHS training programme have? Explain each of them.
5. What are the difficulties of current safety training and teaching?

科技英语翻译技巧——词义的选择和引申

1. 词义的选择

在英语中，一个词语往往有多种含义、多种词类，因此在科技英语翻译中选择出一个符合全文的词义与词类尤为重要。词义的选择通常有如下几种原则：

（1）根据词类选择词义

Like charges repel, unlike charges attract.

同性电荷相斥，异性电荷间相吸。(like 译为“相同的”)

Like a liquid a gas has no shape, but unlike a liquid it will expand and fill any container it is put in.

气体和液体一样没有形状，但又不同于液体，气体会扩散并充满任何盛放它的容器。(like 译为“像”)

（2）根据不同专业的特点选择词义

Just as the sun is the central *body* of the solar system, so the nucleus is the core of the atom.

正像太阳是太阳系的中心天体一样，原子核是原子的核心。(body 译为“天体”)

Recent space travels have shown that the *body* needs special exercise in a spaceship to suit the weightless conditions.

最近的宇宙飞行表明，人体需要在宇宙飞船中进行专门训练才能适应失重情况。(body 译为

"人体")

(3) 根据不同的语境选择词义

In *developing* the design, we must consider the feasibility of processing.

在进行设计时，必须考虑加工的可能性。(develop 译为"进行")

Land animals are believed to have *developed* from sea animals.

人们认为陆地动物是由海洋动物进化而来的。(develop 译为"进化")

(4) 根据英汉的搭配习惯选择词义

Everything in our *physical* world is continually changing.

我们物质世界的一切都在不断地变化着。(physical 译为"物质")

Relations between *physical* quantities are usually given in the form of equations.

物理量之间的关系通常用公式表示。(physical 译为"物理")

(5) 注意在逻辑关系中选择词义

It is impossible to predict in detail the shape and *mechanism* of the robot slave. It might carry its computer and response *mechanism* around with it and also its source of power.

现在详细地预言这种机器人的形状和机理是不可能的。它可能自身带有计算机和反应装置及能源。(mechanism 分别译为"机理""装置")

(6) 根据名词的"数"选择词义

The *time* will come when atomic energy is widely used in industry.

原子能在工业上广泛应用的时候将会到来。(time 译为"时候")

The thermal conductivity of metals is as much as several hundred *times* that of glass.

金属的导热率比玻璃高数百倍。(times 译为"倍")

2. 词义的引申

英语和汉语在表达方式上存在很大的差异，翻译时不能只根据原句的结构和形式机械地翻译，要根据上下文的逻辑关系，将词义引申，恰当地表达出来。引申可以是单词或词组，也可以是整个句子。

(1) 具体化

具体化就是将原文中较为抽象、笼统、概括性的词语引申为有具体含义的词语。

While this restriction on the size of the circuit holds, *the law is valid.*

只要电路尺寸符合上述的限制，这条定律就适用于该电路。

Other *things* being equal, iron heats up faster than aluminium.

当其他条件相同时，铁比铝热得快。

(2) 抽象化

抽象化就是把原文中比较具体、形象化的词引申为比较概括型和抽象化的词语。

There are three steps which must be taken before we *graduate from* the integrated circuit technology.

我们要完全掌握集成电路技术，还必须经过三个阶段。(不译为"毕业于")

The *shortest distance* between raw material and a finished part is casting.

铸造是把原来材料加工成成品的最简便方法。(不译为"最短的距离")

(3) 专业化

专业化就是在专业性的文章中，某些生活中常用的词汇包含了专业的概念，因此将这些词根据专业而引申出词义。

The power dissipated by the filament literally *boils* electrons from the surface of the filament.

灯丝所消耗的功率完全用于从灯丝表面发射电子。(将 boil 引申为“发射”)

The effect of the “load current” can be *duplicated* by connecting a load resistor across the rectifier output and chassis.

负载电流所产生的影响可以等效为在整流器输出端与机壳之间接上一个负载电阻。(将 duplicated引申为“等效”)

(4) 搭配

词的搭配就是在遇到动词、名词、形容词以及与名词的搭配关系时，根据汉语的搭配习惯，为适应名词而把动词或形容词的词义做引申处理。

Threads may be developed on the milling machine by use of thread milling cutters.

使用螺纹铣刀可以在铣床上加工螺纹。(不译为“发展螺纹”)

Such particles are far too tiny to be seen with *the strongest microscope*.

这种粒子实在太小，即使用最高倍的显微镜也看不出来。(不译为“最强的显微镜”)

Putting Safety Training Online

Erike Young faced a tall task when seeking to deliver mandated safety training to 50,000 employees on 10 different campuses across the University of California system. Young, the system's director of environment, health and safety, had a logistics challenge in ensuring that users could access a laboratory safety training course in a timely, cost-effective way. He had to satisfy a tough audience, too.

Some of the trainees would be professors with doctorates and Nobel Prizes-an audience that could likely teach the course itself and that had little time for training amid research and teaching demands. So when Young opted for a web-based training delivery format, he knew from interviewing a committee of administrative and academic leaders that the e-learning had to be instructionally sound, engaging and effective. After all, this was a culture that still championed instructor-led education as the preferred teaching method.

“The message from the committee was, ‘Don't just tell us to take the training because it's required by regulation,’” says Young, who is based in Oakland. “If our employees had to take ***mandatory*** training, they really wanted them to learn something from it, not just have it be a ‘check-the-box’ exercise.”

Working with ***vendor*** Vivid Learning Systems of Pasco, Wash., Young developed a custom two-hour online lab safety fundamentals course that addresses key regulations from the US Occupational Safety and Health Administration (OSHA). The course has interactive exercises and feedback elements to keep participants engaged and a pre-training assessment that lets employees test out of content they can show they've already mastered. To stress the importance of the training, a short video from university president Mark Yudof kicks off the course.

The university invested about $ 100,000 in course development, but it expects to achieve delivery cost savings over time using online instead of instructor-led training methods.

“We didn't just want to deliver PowerPoint slides with voice-overs that satisfied certain regulations but made people tune out. It was essential that our employees' first experience with online safety training be positive," Young says.

1. More Flexibility

While trainers moved quickly in the past to migrate other training subjects to online delivery, safety training has been slower to join the e-learning party. That's due in part to OSHA, the US Department of Transportation and other ***regulatory*** bodies for work-place hazard training, which require a "hands-on" learning component and ample opportunity for feedback during training.

Improvements in instructional design and growth of Internet ***bandwidth***, however, have enabled developers to build more interactivity, gaming techniques and feedback mechanisms into online safety training, be it authored in-house or purchased off the shelf from training vendors. Delivering safety training through e-learning appeals to HR and training leaders on a number of fronts:

- It ensures that all employees receive the same training in the same way.
- It can be accessed around the clock by employees working from remote locations.
- It provides more-efficient reporting upon completion of training classes than instructor-led options.

Those records are essential to meet regulatory requirements. In other cases, proof of safety training completion is needed to ***grant*** workers access to certain areas of work environments.

"More companies are turning to blended, flexible or mobile versions of safety training," says Tess Taylor, PHR, who writes about workplace safety training issues for the blog HR writer. "That's partly because more employees are working remotely or in flex arrangements, but also because it can be more cost-effective from a delivery standpoint to conduct training that way."

2. Find the Right Tool

When Jack Hawkins, director of risk control, started working at the Coca Cola Bottling Cos. in Charlotte, N. C., safety training was conducted through instructor-led courses and CD-based video learning. While those offline formats met regulatory objectives, they fell short of other training goals. "We often start the day at 5 a. m., and—for younger employees especially—having to come in and sit in the classroom at that hour for safety training wasn't conducive to motivation," Hawkins says.

The company has since migrated to e-learning for safety education. Employees can get online instruction from remote locations or from dedicated computer stations at 50 product distribution centers in nine states. Content is accessed by employees via a software-as-a-service model from vendor UL Pure Safety in Franklin, Tennessee, an option Hawkins chose for its user-friendliness and instructionally sound content.

"We didn't want our people to have to be computer scientists to log onto the site, find their safety lessons and take the training," Hawkins says.

Most of the company's safety training is conducted around driver safety and workplace injury and accident prevention to ***comply with*** requirements from the US Department of Transportation and OSHA. Hawkins says the learning management system supporting the online training has been a benefit in tracking compliance.

With a new ability to monitor companywide training completion percentages, Hawkins could see early in training ***implementation*** that at the end of a given month, only 60 percent of employees had completed assigned safety lessons.

To boost that rate, Hawkins began publicizing training completion "scorecards" by location and department. Within months, spurred by the scorecards and supervisors' encouragement, employees had achieved 80 percent compliance. Now, "We are close to 100 percent compliance for the 65,000 safety lessons our employee population takes over the course of a year," Hawkins says. Coca-Cola Bottling structures the training in a way that keeps workplace safety issues top of mind for employees. "We don't want people to be able to simply log onto the site in January and take their 12 safety lessons and he done," Hawkins says. "The system allows us to assign one or two lessons a month across the whole organization, so people can't work ahead."

Keeping e-learning ***engaging for*** employees is vital, Hawkins says. Most training sessions are kept for 20 to 30 minutes, including testing time, and each has some form of interactive exercise to hold learners' attention.

E-learning has had a positive impact on workplace safety performance, Hawkins says. According to a company-developed "risk dashboard" that details injury and vehicle accident rates for all operating units, Coca-Cola Bottling has had fewer injuries and vehicle accidents each year for the past five years, he says. "We've reduced the overall frequency of vehicle injuries and accidents about 50 percent in that time, and, as a result, the cost of those losses is down 50 percent on an annualized basis," Hawkins says.

3. Considerations

While cost and content standardization factors weigh heavily in decisions to deploy online safety training, HR and training leaders usually also want assurances that the safety messages they've approved are delivered uniformly to all employees and can be updated and delivered across dispersed audiences. E-learning can satisfy all these demands.

Marc VanderWal, director of global environment, health and safety for Graham Packaging Co. in York, Pa., was looking for a system that had content that would stay current with shifting industry regulations, have an engaging instructional design and be available in multiple languages. In evaluating potential vendors, be wanted a competitive price, too.

VanderWal chose safety training content from Dupont Sustainable Solutions in Virginia Beach, Va., to deliver to 6,500 employees working in 14 countries. Graham pays "$20 to $25 per employee per year" for access to 10 safety training courses.

In the past, Graham's plants purchased or developed safety training courses on their own. But the process was ***inconsistent***, VanderWal says, and HR managers in some plants were using content that didn't comply with regulatory standards. That issue was resolved by centralizing delivery under Graham University, the organization's education arm. Under that structure, the vendor's learning management system tracks training completion rates, sends out e-mail reminders when refresher training is needed and handles other administrative tasks. Previously, HR managers in each plant tracked and reported on training, typically using Excel spreadsheets.

Graham's operating plants do retain some autonomy, however. They're given the choice of delivering training via the Web, DVD-based formats or instructor-led options. "Having different delivery methods gives plant leaders some options in customizing learning in ways they think work best for their distinct work cultures," VanderWal says.

At Key Technology, a maker of process automated systems in Walla Walla, Wash., using e-learning for safety education has proved to be a good fit for 20 field engineers throughout the United States, says John Kadinger, a market manager who oversees staff training.

"The engineers work remotely, so they can log on and complete training from home offices," Kadinger says. "It's more cost-effective and convenient than flying them into headquarters for training or hiring outside instructors to teach courses at regional locations."

4. High Standards

Even though cost and ease of access are factors in choosing an online safety program, some safety leaders say the biggest issue to consider is how training affects employees' behavior on the job. "You can do things very cheaply, but if it's not going to affect behavior, you will pay for it in other ways," says the University of California's Young.

Another key criterion in evaluating off-the-shelf courseware is ensuring that training is kept current with ever changing OSHA or other industry regulations. VanderWal says that was a top priority for his organization. "If there is any liability regarding gaps in our safety training, we know our vendor is constantly surfing regulations to make sure the content is updated," he says. "That is a big value-add for us."

Given the serious, often life-and-death nature of safety lapses in the workplace, the stakes are becoming even higher for HR and training leaders to deliver online training that meets the highest standard of instructional quality.

New Words and Expressions

mandatory ['mændət(ə)ri] *adj.* 强制的，法定的，义务的
vendor ['vendə] *n.* 〈正式〉供应商，小贩
regulatory ['regjʊlətəri] *adj.* 管理的，调整的，监管的
bandwidth ['bændwɪtθ] *n.* 带宽，频宽，带宽值
grant [grɑːnt] *v.* 允许，同意
comply with 服从，遵从，顺应，照办
implementation [ɪmplɪmen'teɪʃ(ə)n] *n.* 执行，落实
engage for 保证，对……负责
inconsistent [ɪnkən'sɪst(ə)nt] *adj.* 不一致，相矛盾，不符合，反复无常的

Unit Seven

The Practice of Mine Ventilation Engineering

The practice of ventilation is continually evolving with new technological advances developed in the mining industry. In recent years the advances in diesel engine technologies, ventilation modeling software, and ventilation management capacities have redefined the historical methods used to evaluate systems. The advances re-evaluate previous methods used to calculate the airflow requirements for the dilution of diesel exhaust fumes. Modeling software has become an integral part of planning and developing ventilation systems in partnership with graphical mine design software packages to generate realistic representations of the mine. Significant advances in ventilation control strategies through remote sensors and monitoring capabilities have been developed to results in cost savings. Though there has been much advancement in ***mine ventilation*** technology, the practices and basic ventilation principals enacted through the ventilation engineer cannot be placated with technological advances only.

Ventilation of any underground mine is a process of identifying the potential hazards from a mining system and implementing a ventilation system that ensures the health and safety of the mine worker. To this end, a number of new advances have been made to benefit the engineer in achieving this goal. To understand the issues with providing a healthy environment, one must first evaluate the hazards.

In mines, we rarely have a choice in deciding the properties of the ore or coal body, the surface climate, the inflow of strata water and/or gas, and the physical and chemical properties of the rock. What we do have control over are the design factors, such as, methods of mining, layout of airways, rate of rock ***fragmentation***, and types of vehicles to be used. However, it is often the case that much of the design will be driven by production considerations and not by ventilation concerns. Ventilation is often thought of as an expense and not a means to a profitable mine. This thought, however, is not justified since it is obvious that a healthy workplace results in greater productivity. This is challenging to prove in practice since it is difficult to quantify the difference in productivity between a poor and good ventilated mine.

In today's modern mines, the engineer needs to have solid understanding of legislative requirements and "best" engineering practice. Common parameters legislated can include gas and dust concentrations,

thermal conditions, Diesel Particulate Matter (DPM), and minimum airflow rates. The engineer needs to understand these requirements when designing a ventilation system. Good practice means incorporating a ventilation system that fulfills the legislative requirements even if there is no mine inspectorate to ensure the system is in compliance. Being complacent in the practice of mine ventilation can have severe, and possibly deadly, consequences.

1. Current Trends in Mine Ventilation

In metal mines there is a trend for larger diesel equipment, increased production pressures resulting in a larger number of operating faces, the use of ***series ventilation*** to minimize airflow, increased regulations to lower ***respirable*** particulate dust, increased refrigeration requirements as mines go deeper, use of haulage ramp air as an intake or exhaust from sublevels, and an increase in electrical power costs that can drive designs to a minimal ventilation system.

In coal mines the trends include, improved real time communication and tracking systems for effective escape and/or refuge planning, applying ***inert*** gas injection to sealed gob areas to reduce explosive atmospheres, improved rock dusting application to minimize explosive dust, increased legislative requirements to control silica and coal dust exposers, and use of underground booster fans to enhance ventilation to working areas.

2. Ventilation on Demand

The concept of Ventilation on Demand (VOD) is to apply airflow to only the working areas of the mine while minimizing airflow to remaining areas. This concept is typically applied to metal/non-metal mines and not coal mines. The system can be as simple as one that turns on ventilation to a zone regardless of the work activity or a relatively complicated one that controls the flow based on air quality sensors. The later system usually requires fan motors on variable speed drives (or variable frequency drives-VFD), air gas sensors (e. g. carbon monoxide, oxygen, ***nitrogen*** oxides, etc.), airflow sensors, regulator and fan control systems, and equipment and personnel tagging systems. The concept is to provide airflow as needed during the mining cycle. For example, an LHD entering a stope would require a specific airflow rate. This airflow rate would be predetermined for the LHD. A regulator or fan would be opened to provide this airflow. The tagging system would identify the location of the LHD to ensure the flow is constant during its operation in the area. Air quality sensors monitor the air condition during the mining cycle. When the LHD leaves the area, these sensors will maintain the airflow rate until such a time as the air quality is acceptable and the regulator or fan can be turned down or off. This logic would apply to any operating equipment in the mine and for personnel. In addition, the primary fans would also have VFD control.

Some calculations have shown that a fully automated VOD system can have an electrical power savings of up to 50% over a conventional mine ventilation system. The use of VOD for coal mines is far more challenging since many governments legislate minimum airflow quantities at strategic locations. Varying the flow could have serious consequences if the sensors or control systems are not operating correctly.

Tools available include computer software for ventilation planning and implementation of monitoring systems in the mine to evaluate fans, airflow, temperature, gasses, and other parameters in real time.

3. Ventilation Monitoring Systems

In some countries monitoring of certain fan operating conditions and/or air quality is mandatory. Continuous fan static pressure measurements are required at the US coal mines. In addition, monitoring of explosive gases in sealed gob and in entries is required. Other parameters measured can include carbon monoxide, carbon dioxide, oxygen, and methane. For metal mines, the gases measured are similar to above but may include sulfide gases and nitrates of oxide. Additionally, on conveyors smoke sensors are often included in the monitoring program.

Other parameters monitored are airflow at strategic underground locations and through fan systems, door and regulator positions, and air temperatures. Other parameters can include DPM and dust continuous monitors. Subsystems can include water flow and temperature for heat exchangers (water sprays) and fan data such as on/off and operating position if the motor is equipped with a VFD.

These systems allow for real time evaluation of the underground environment and can be used in a VOD system to provide ventilation flow control to minimize fan power costs.

4. Power Savings in Ventilation

At many underground mines, operating the ventilation system continuously is a significant percentage of the power used. Up to 70% of the mine power can be devoted to primary fans, air chillers and other ventilation components. Power savings are achieved by reducing flow during times when no activities are occurring in that area of the mine and no contaminants are accumulating. The VOD section above describes this approach. Other areas where significant power savings can be achieved are with air cooling systems. Some refrigeration plants are currently being designed to manufacture ice during the evening and using this as water for the surface spray chambers. By producing ice during low electrical demand periods (and at a lower price) and melting and using it in the surface heat exchanger during high electrical demand periods has resulted in significant power savings at some South African mines.

Ventilation engineers need to be very cognizant in the design of ventilation and air conditioning systems to ensure efficient and cost effective systems are installed. The use of software to optimize ventilation and refrigeration systems is critical to optimizing these systems.

5. Current Ventilation Concepts for Coal Mining

The ventilation of coal mines and maintaining a safe atmosphere for the workers depends upon many factors outside of simply ventilating for equipment and blast fume clearances. Coal mine ventilation is primarily concerned with controlling explosive gas and dust, oxygen ***deficiency***, ***spontaneous combustion*** and dust control (pneumoconiosis). Because explosive gas can be generated separately from the mining process, coal mines cannot and should not rely on VOD systems to automatically adjust fans and/or flow in the mine. Changing ventilation in a coal mine should be a manually controlled operation.

In many developed countries, government agencies mandate specific ventilation levels and conditions throughout a coal mine. In sealed areas of a coal mine or in areas behind a longwall the possibility of a buildup of explosive gas mixtures is of primary concern. Therefore, ensuring an inert atmosphere in gob areas is a primary concern. There are two thoughts to this process. Either the gob area is maintained at a very high explosive gas level (which is not explosive) or the area is kept well below the explosive gas level. The hazard with the first approach is that any gas leaking from the gob area to the active mine will transition through the explosive range. Such leakage often happens during periods when the surface barometric pressure is falling (creating a differential pressure between the sealed area and the mine airways). Because of this, more coal companies are ensuring the gob areas are kept well below the explosive gas levels by using bleeder ventilation systems. These systems intentionally apply intake air behind the mining areas to extract the gas and send it to the return. If these systems are not capable of maintaining the gas in a safe, low concentration then either draining the gas with drainage systems or injecting the area with an inert gas such as nitrogen may be necessary.

In coal mine ventilation design, traditional conveyor systems have been placed in neutral airways to prevent the air used to ventilate these systems from actively ventilating the working faces. However, with increasing production rates leading to higher gas liberation and elevated airflow, the conveyor entries have been used to provide fresh air to the working faces in conjunction with the fresh air entries. In order to use the conveyor air at the production face, the ventilation system requires bulkheads to separate the fresh air and conveyor air. This provides a dedicated fresh air escape way to be maintained from the working faces and a parallel route for the additional air capacity to reach the working faces. In addition to the physical separation, gas/smoke sensors are installed along the length of the conveyor systems to ensure early detection of conveyor fires and allow for rapid notification to underground personnel.

As new ventilation technology emerges we must not lose focus on the basic principles. The use of ventilation modeling software represents a significant time saving tool and can greatly assist the ventilation engineer in developing complete and thorough designs, allowing for the rapid development of numerous permeation and design options. However, the ventilation engineer should never lose sight of first principles. The tools available to the ventilation engineer are only as good as the project inputs developed by the engineer.

Technological advances in monitoring and control systems have made the implementation of VOD systems a reality. However, they must be properly designed such that safety is not sacrificed in the search of increasing efficiency and decreasing power and infrastructure costs. Developing and installing monitoring systems has increased the level of safety by allowing continuous monitoring of gasses and temperatures throughout the mining areas. This allows the mine to evacuate or receive notification when adverse or dangerous conditions are encountered. The continued advances in ventilation technology will help to elevate the health and safety of the miners, as long as the ventilation engineers do not lose perspective of the founding principles of ventilation.

New Words and Expressions

mine ventilation 矿井通风

fragmentation [frægmen'teɪʃ (ə) n] *n.* 破碎，[生]（染色体）断裂

thermal ['θɜːm (ə) l] *adj.* 热的，热量的，保暖的，防寒的
n. 上升的热气流，保暖内衣裤

series ventilation 串联通风

respirable ['resp (ə) rəb (ə) l] *adj.* 可呼吸的

inert [ɪ'nɜːt] *adj.* 无活动能力的，无行动力的，惰性的，不活泼的

nitrogen ['naɪtrədʒ (ə) n] *n.* 氮，氮气

deficiency [dɪ'fɪʃ (ə) nsi]　　n. 缺乏，不足，缺陷，缺少
spontaneous combustion　　自燃

Exercises

1. What is ventilation? What is main ventilation?
2. Whatis the challenging for development of main ventilation?
3. What do engineers need to do when designing a ventilation system in today's model mines?
4. What is the good practice means in mine ventilation?
5. Explain the current ventilation concepts for coal mining.

科技英语翻译技巧——词类转化

英语和汉语在语言结构和表达方式上存在着差异，因此在翻译时，为了更符合汉语的表达习惯，使译句更加准确通顺，常常需要采用词类转换的方式，即将某一词类转化成另一词类翻译。

以下几种情况，需要考虑词类转换：

1）在英语中的某一类词在汉语中相当于另一类词时。

2）当某一个词不能用本来的词类确切表达该词的含义时。

3）当某一个词按照本来的词性翻译，会使译句不通顺时。

1. 转译为名词

（1）动词转译为名词

Boiling point *is defined* as the temperature at which the vapor pressure is equal to that of the atmosphere.

沸点的定义就是蒸气压力等于大气压时的温度。

（2）代词转译为名词

The most common acceleration is *that* of freely falling bodies.

最普通的加速度是自由落体加速度。

（3）形容词转译为名词

TV is *different* from radio in that it sends and receives a picture.

电视和收音机的区别就在于电视发送和接收的是图像。

（4）副词转译为名词

Lithium, sodium, potassium, and copper have one electron in the outer shell and are *chemically* similar.

锂、钠、钾和铜在最外电子层只有一个电子，因而其化学性质相似。

2. 转译为动词

（1）名词转译为动词

One of the principal rules of the lathe maintenance is timely *lubrication* of all friction surfaces.

车床维护的主要规律之一就是及时润滑所有摩擦面。

（2）形容词转译为动词

The amount of work is *dependent on* the applied force and the distance the body is moved.

功的大小由所施的力与物体移动的距离确定。

（3）副词转译为动词

If one generator is out of order, the other will produce electricity *instead*.

如果一台发电机发生故障，另一台便代替发电。

（4）介词转译为动词

Also, finned heat sinks were used at the lamp terminals which were cooled *by* a forced draught of compressed air.

还有一套散热装置用于灯的末端，它采用压缩空气进行强制通风冷却。

3. 转译为形容词

（1）名词转译为形容词

Much less is connected with the separation of genera, and there is considerable *uniformity* of opinion as to the delimitation of families.

这与属的划分关系不大，而在科的划分上观点是相当一致的。

（2）副词转译为形容词

Miniature gas detector is *chiefly* featured by small size, light weight, complete functions and long continuous working time.

袖珍式瓦斯检测器的主要特点是体积小、重量轻、功能齐全、连续工作时间长。

（3）动词转译为形容词

The range of the spectrum in which heat is *radiated* mostly lies within the infrared portion.

辐射热的光谱段大部分位于红外区。

4. 转译为副词

（1）形容词转译为副词

The potentials of both the grid and plate are effective *in* controlling plate *current*.

栅极和板极两者的电位有效地控制着板极电流。

（2）名词转译为副词

In human society activity in production develops *step by step* from a lower to higher level.

人类社会的生产活动，由低级向高级逐步发展。

（3）动词转译为副词

The molecules continue to stay close together, but do not *continue* to retain a regular fixed arrangement.

分子仍然紧密地聚集在一起，但不再继续保持有规则的固定排列形式。

Challenges to Developing Methane Biofiltration for Coal Mine Ventilation Air

Coal mine **methane** is a significant green-house gas source as well as a potential lost energy resource if not effectively used. In recent years, Mine Ventilation Air (MVA) capture and use has become a key element of research and development due to comparatively larger methane emissions by MVA than other coal mine sources. Technologies have been evaluated to treat the low methane concentrations in MVA such as thermal-based technologies or processing by ***biofiltration***. This review initially considers the techniques available for treating the low methane concentrations encountered in MVA, after which it focuses on developments in biofiltration systems. Biofiltration represents a simple, energy-efficient, and cheap alternative to oxidize methane from MVA. Major factors influencing biofilter performance along with knowledge gaps in relation to its application to MVA are identified and discussed.

1. Coal Mine Ventilation Air

CH_4 in coal beds is formed during the coal creation process and resides within the coal seam and adjacent rock strata. CMM encompasses all CH_4 released prior to, during, and after mining operations. If the mine is "gassy", it is possible to crack the coal seam prior to mining and capture the gas at reasonably high concentration [60%–95 % CH_4 (v/v) in air]. It is also possible to concentrate CH_4 gas from worked areas of the mine to 30%–95 %. Such CH_4 concentrations can be thermally oxidized using a conventional gas ***turbine*** to generate electricity. However, most coal mines are not gassy, releasing only fugitive emissions.

Underground mine operators employ ventilation to dilute combustible gases in the mine. Ventilation systems consist of inlet and exhaust shafts and powerful fans, which transport large volumes of air through the mine sites to sustain a safe working environment, usually defined as <2 % CH_4 (v/v) in air, depending on the jurisdiction. The CH_4 concentration in exhausted ventilation air is usually maintained at around 0.2% to 1.0 % (v/v) in gassy mines and 0.1 % (v/v) in non-gassy mines to keep it well below the ignition limits [typically around 5%–15 % (v/v) CH_4 in air]. Only a few technologies are currently available to capture these low CH_4 concentrations in MVA as an energy source. In fact, in most cases, the diluted CH_4 (in air) is directly released to the atmosphere without any attempts to capture it. The difficulty with its effective use derives from the high MVA flow rate (50 – 500 m^3/s), alongside the variably low methane concentration [0.1%–1 % (v/v) in air].

Apart from a low CH_4 concentration, MVA can contain other components which may hinder operations of methane removal technologies. Determined the MVA ***composition*** from four coal mines located within New South Wales and Queensland in Australia. All the assessed MVAs contained dust that consisted of micron-sized coal and stone particles, which may increase maintenance costs due to particle deposition. In fact, it has been reported that the use of internal combustion engines at Appin Colliery (New South Wales, Australia) for CH_4 removal from MVA was discontinued due to cost inefficiencies and frequent cleaning required to remove dust particles. Low levels (<1 ppm in air) of carbon monoxide (CO), hydrogen sulfide (H_2S), and ***sulfur*** dioxide (SO_2) have also been reported.

2. Ventilation Air Methane Regenerative after Burner

Cork and Peet developed ventilation air methane regenerative after burner (VAM-RAB) technology to remove CH_4 from coal MVA. A pilot-scale setup has been commissioned and run at Centennial Coal's Bloomfield site (New South Wales, Australia). According to Myors, the pilot-scale VAM-RAB is able to completely oxidize VAM (99.98 % CH_4 removal efficiency) at 1 m^3/s ventilation air flow rate and inlet CH_4

concentration in the range 0.35%–1.1% (v/v) in air. Full-scale implementation is currently planned at Mandalong site (New South Wales, Australia) to treat up to 300 m^3/s gas flow rate from MVA with capacity to reduce 650,000 tonnes CO_2 equivalent per annum.

The disadvantages of this technology include its large capital and operating cost due to expensive installation and operation of the reactors, as well as complicated safety equipment (e.g., back-flash protection) and the large area requirement of 36 units of VAM-RAB required to treat large MVA flow rates (300 m^3/s).

3. Biofiltration

In the context of air pollution, biofiltration is defined as a system that employs microbes to catalytically degrade pollutants from a contaminated exhaust stream. For instance, CH_4 from an exhaust stream can be converted into biomass, carbon dioxide, and water using methanotrophic microbes. Unlike conventional CH_4 removal technology, biofiltration offers great advantages since it is relatively cheap to build and operate as it works at ambient temperature and pressure. The ambient conditions and lack of electrical components also decrease the need for complicated safety devices. Biofiltration may tolerate variable input flows and concentrations and is easy to install and mobile which suits a potential MVA application. Biofiltration systems have already been shown to operate at high efficiency when used for high flow waste gases with low concentrations of pollutants [≤1 % (v/v) in air].

Biofiltration is a green process, which does not generate any hazardous pollutants such as oxides of nitrogen (NO_x), particulate matter, SO_2, or CO in the exhaust gas. The biofiltration system typically consists of organic or inorganic packing materials as the growth surface for the microbes. As a rich supply of macronutrients, organic beds are used to sustain biomass growth. Biofiltration has been used over the 20th century to treat sewage and other odoriferous, waterborne wastes. Most European countries have used bioreactors to treat contaminated air for 60 years.

There are several biofiltration systems that have been utilized for CH_4 elimination, namely, conventional biofilter, biotrickling filter, bioscrubber, membrane bioreactor, and two-liquid-phase bioreactor. Several laboratory-scale biofilter studies utilized a biotrickling filter, which is a style of biofilter where nutrients are continuously fed into the top of the unit. However, this paper will focus on the conventional biofilter for coal MVA application since there is no continuous feed of a liquid phase which makes the technology easy to install, mobile, and low in cost. The mechanism of biological CH_4 oxidation by methanotrophs (methane-oxidizing microbes) and the design of a conventional biofilter will be discussed in the next subsections.

4. Mechanism of Biological CH_4 Oxidation by Methanotrophs

Methanotrophs are a physiological group of bacteria with the ability to utilize CH_4 as a sole carbon and energy source. There are several intracellular steps during CH_4 utilization, as shown in Scheme 1. The first reaction step is the oxidation of CH_4 to methanol (CH_3OH) catalyzed by the ***enzyme*** CH_4 monooxygenase (MMO); this is followed by oxidation of CH_3OH to formaldehyde (HCHO) catalyzed by methanol dehydrogenase (MDH). Formaldehyde then serves as a ***substrate*** in either a dissimilatory pathway via formate (HCOOH) to CO_2 for energy generation or various assimilatory pathways leading to the synthesis of cell components, essential for the growth of methanotrophs.

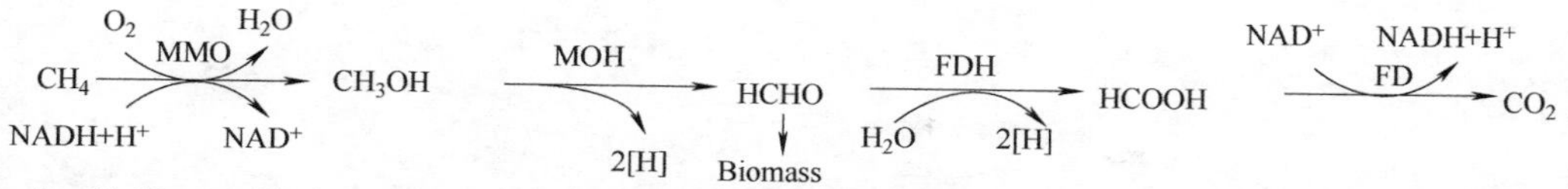

$$CH_4 \xrightarrow[NADH+H^+ \quad NAD^+]{O_2 \quad MMO \quad H_2O} CH_3OH \xrightarrow[2[H]]{MOH} HCHO \xrightarrow[H_2O \quad 2[H]]{FDH} HCOOH \xrightarrow{NAD^+ \quad FD \quad NADH+H^+} CO_2$$

HCHO ↓ Biomass

Scheme 1 Methane oxidation pathway of methnotrophic bacteria (Hanson and Hanson 1996; Razumovsky et al. 2008). NAD^+ means nicotinamide adenine dinucleotide oxidized form, NADH means nicotinamide adenine dinucleotide reduced form, MMO means methane monooxy genase, MDH means methanol dehydrogenase, FDH means formaldehyde dehydrogenase, FD means formate dehydrogenase.

Methanotrophs can be classified as type Ⅰ, type Ⅱ, and type X based on their cell morphology, assimilatory pathway, growth temperature, nitrogen fixation, and ***membrane*** arrangement. The genera Methylomonas, Methylomicrobium, Methylobacter, Methylocaldum, Methylophaga, Methylosarcina, Methylothermus, Methylohalobius, and Methylosphaera belong to type I. These genera ***assimilate*** HCHO (an intermediate of CH_4 oxidation) by the ribulose monophosphate pathway, and their cellular membranes consist of fatty acids with 16 or sometimes 14 atoms of carbon. Examples for type Ⅱ methanotrophs are the genera Methylocystis, Methylocella, Methylocapsa, and Methylosinus. Type Ⅱ methanotrophs use the serine pathway to assimilate HCHO, and their cellular membranes are mainly made up of fatty acids with 18 carbons arranged around the cell ***periphery*** (Borjesson et al. 1998; Hanson and Hanson 1996; Nikiema et al. 2005). The genus Methylococcus belongs to type X methanotrophs, which combine the properties of type I and type Ⅱ. Their cellular membranes contain fatty acids with 16 carbons with HCHO assimilation via the ribulose monophosphate and the serine pathway.

5. Challenges in Developing CH_4 Biofiltration to Eliminate CH_4 from Coal MVA

Biofiltration is not as ***prone to*** high capital and operating cost issues (compared with other potential technologies) but may be sensitive to several key operating ***parameters***, such as variations in operating conditions, for instance temperature, moisture content, pH, and nutrient content. A further hindrance is the poor solubility of CH_4 in the aqueous phase. The low CH_4 solubility necessitates a higher residence time, which subsequently demands an increase in bioreactor size and capital cost. Several attempts have been made to improve CH_4 biofiltration performance, which are detailed in the ensuing paragraphs.

New Words and Expressions

methane [ˈmiːθeɪn] n. 甲烷，沼气
biofiltration [bɪəfɪltˈreɪʃn] n. 生物过滤，渗入，渗透
turbine [ˈtɜːbaɪn] n. 汽轮机，涡轮机
composition [ˌkɒmpəˈzɪʃn] n. 成分，构成，作文
sulfur [ˈsʌlfə] n. 硫黄
in the context of 在……情况下，在……背景下
enzyme [ˈenzaɪm] n. 酶
substrate [ˈsʌbstreɪt] n. 底物，基底，基层，底层
membrane [ˈmembreɪn] n. （身体内的）膜，（植物的）细胞膜
assimilate [əˈsɪmɪleɪt] v. 吸收，消化，透彻理解，（使）同化
periphery [pəˈrɪf (ə) ri] n. 外围，边缘，周围，次要部分
prone to 易于，倾向于
parameter [pəˈræmɪtə] n. 参数，参词，参项，参量

Unit Eight

Classification of Fires and Fire Hazard Properties

A major safety concern in industrial plants is the occurrence of fires and explosions. Hazardous locations are areas where flammable liquids, gases, or ***vapors*** or combustible dust exist in sufficient quantities to produce an explosion or fire. In hazardous locations, specially designed equipment and special installation techniques must be used to protect against the explosive and flammable potential of these substances.

Flammable materials are substances that can ignite easily and burn rapidly. Flammable substances can be divided into three subgroups:

- flammable gas;
- flammable liquids;
- flammable solids.

1. Classification of Fires

Fire is the rapid oxidation of any combustible material. It is a chemical reaction involving fuel, heat, and oxygen. These three elements, commonly referred to as the fire triangle, in the right proportions will always produce a fire. Remove any one side of the triangle and the fire will be ***extinguished***.

It should be noted that every flammable gas or vapor has special lower and upper flammability limits. If the substance or concentration in the oxidizer is either below a special value (lower flammability limit) or above a special value (upper flammability limit), ignition might occur; however, a flame will not ***propagate***. This can be exploited by diluting the flammable substances with air or preventing the ingress of air/oxygen. The latter option is ruled out in environments where people work regularly and is feasible only in a chemical plant where there are no human beings.

If a flammable gas or vapor cloud is released and ignited, all the material may be consumed in one explosion. If the flammable gas or vapor cloud is not ignited, convection and diffusion will eventually disperse the flammable cloud, the immediate danger passes, and the particular fuel source is lost.

The most common types of reaction are between flammable gases, vapors, or dust with oxygen contained in the surrounding air.

Figure 8-1 shows the fire triangle. It is commonly used as a model to understand how a fire starts and how it can be prevented. The presence of these three elements makes up the sides of the ignition triangle. If any one of the three elements is missing, an explosion will not occur. All three elements must exist simultaneously for an explosion to occur.

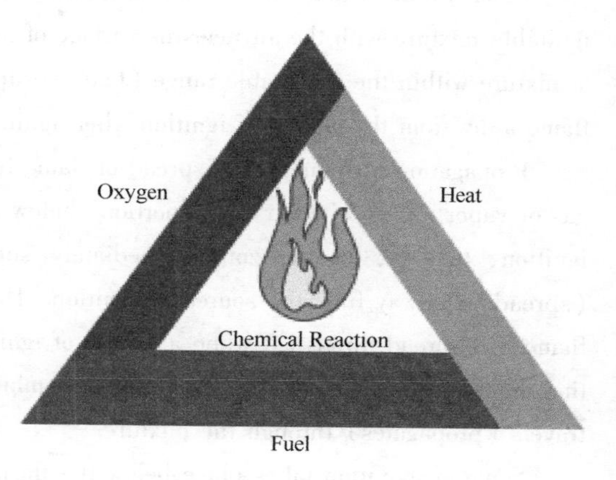

Figure 8-1 Fire Triangle

(1) Oxidizer

The oxidizer ***referred to*** in all common hazardous location standards and explosion-proof equipment is air at normal atmospheric conditions. The oxygen in the air is only enough for the combustion of a certain quantity of flammable material. Air must be present in sufficient volume to propagate a flame before the air—fuel mixture becomes a hazard.

When the amount of available atmospheric oxygen is more or less in equilibrium with the quantity of flammable material, the effect of an explosion—both temperature and pressure—is most violent. If the quantity of flammable material is too small, combustion will spread with difficulty or cease altogether. The same applies if the quantity of flammable material is too great for the available oxygen.

The presence of an oxygen-enriched atmosphere or a pressurized ***enclosure*** alters the conditions for ignition and dictates the use of special means for prevention and containment of explosions. No means of explosion protection considered safe for atmospheric mixtures should be used in either oxygen-enriched or pressurized situations without careful study.

(2) Ignition Source

The amount of energy required to cause ignition is dependent upon these factors:

- the concentration of the hazardous substance within its specific flammability limits;
- the explosive characteristics of the particular hazardous substance;
- the volume of the location in which the hazardous substance is present.

2. Fire-Hazard Properties

A single fire-hazard property such as flash point or ignition temperature should not be used to describe or appraise the hazard or fire risk of a material, product, assembly, or system under actual fire conditions. The subject fire-hazard properties have been determined under controlled laboratory conditions and may properly be employed to measure or describe the response of materials, products, assemblies, or systems under these conditions.

Properties measured under such conditions may be used as elements of afire risk assessment only when such assessment takes into account all of the factors that are pertinent to the evaluation of the fire hazard of a given situation. Properties of the flammable materials are generally for materials in the pure form and may be different if there are impurities or where there are mixtures of materials.

(1) Flash Point

Flash point of the liquid is the minimum temperature at which it gives off sufficient vapor to form an ignitable mixture with the air near the surface of the liquid within the vessel used. An "ignitable mixture" is a mixture within the flammable range (between upper and lower limits) that is capable of the propagation of flame away from the source of ignition when ignited.

Propagation of flame is the spread of flame from the source of ignition through a flammable mixture. A gas or vapor mixed with air in proportions below the lower limit of flammability may burn at the source of ignition, that is, in the zone immediately surrounding the source of ignition, with out propagating (spreading) away from the source of ignition. However, if the mixture is within the flammable range, the flame will spread through it when a source of ignition is supplied. The use of the term flame propagation is therefore convenient to distinguish between combustion, which takes place only at the source of ignition and travels (propagates) through the mixture.

Some evaporation takes place below the flash point but not in sufficient quantities to form an ignitable mixture. This term applies mostly to flammable and combustible liquids, although there are certain solids, such as camphor and naphthalene, which slowly evaporate or volatilize at ordinary room temperature, or liquids such as benzene that freeze at relatively high temperatures (5.5℃) and therefore have flash points while in the solid state.

The test apparatus used for the measurement of flash point is normally one of two types, of which there are several variants. These are called generally open cup and closed cup flash point testers. For most liquids the flash point determined by the closed cup method is slightly lower (in the region of 5%–10% when measured in degrees Celsius) than that determined by the open cup method.

(2) Ignition Temperature

Ignition temperature of a substance, whether solid, liquid, or gaseous, is the minimum temperature required to initiate or cause self-sustained combustion independently of the heating or heated element and in the absence of any source of ignition.

Ignition temperatures observed under one set of conditions may be changed substantially by a change of conditions. For this reason, ignition temperatures should be looked upon only as approximations Besides, variables known to affect ignition temperatures are percentage composition of the vapor or gas-air mixture, shape and size of the space in which the ignition occurs, rate and duration of heating, and type and reactivity of other materials present in the space in which the ignition occurs.

Some of the variables known to affect ignition temperatures are percent age composition of the vapor or gas-air mixture, shape and size of the space where the ignition occurs, rate and duration of heating, kind and temperature of the ignition source, ***catalytic*** or other effect of materials that may be present, and oxygen concentration.

As there are many differences in ignition temperature test methods, such as size and shape of containers, method of heating, and ignition source, it is not surprising that ignition temperatures are affected by the test method.

(3) Specific Gravity

The specific gravity of a substance is the ratio of the weight of the substance to the weight of the same volume of another substance. (Temperature affects the volume of liquids, and temperature and pressure affect the volume of gases. It is therefore necessary to make corrections for effects of temperature and pressure when making accurate specific gravity determinations.)

Specific gravity refers to the ratio of the weight of a substance to the weight of an equal volume of water.

(4) Relative Vapor Density

The relative vapor density of a material is the mass of a given volume of the material in its gaseous or vapor form compared with the mass of an equal volume of dry air at the same temperature and pressure. It is often calculated as the ratio of the relative molecular mass of the material to the average relative molecular mass of air (the value of the latter being approximately 29).

(5) Boiling Point

The boiling point of a liquid is the temperature of a liquid at which the vapor pressure of the liquid equals the atmospheric pressure. Therefore, the lower the boiling point is, the more volatile and generally the more hazardous the flammable liquid becomes.

Where an accurate boiling point is unavailable for the material in question or for mixtures that do not have a constant boiling point, for purposes of this classification the 10% point of a distillation performed in accordance with ASTM D86 (Standard Method of Test for Distillation of Petroleum Products) may be used as the boiling point of the liquid.

(6) Melting Point

Melting point is the temperature at which a solid of a pure substance changes to a liquid.

(7) Boil-Over

Boil-over is an event in the burning of certain oils in an open top tank when, after a long period of quiescent burning, there is a sudden increase in fire intensity associated with expulsion of burning oil from the tank. Boil-over occurs when the residues from surface burning become more dense than the unburned oil and sink below the surface to form a hot layer, which progresses downward much faster than the regression of the liquid surface. When this hot layer, called a heat wave, reaches water or water in oil emulsion in the bottom of the tank, the water is first superheated and subsequently boils almost explosively, over owing the tank. Oils subject to boil-over consist of components having a wide range of boiling points, including both light ends and viscous residues. These characteristics are present in most crude oils and can be produced in synthetic mixtures.

It should be noted that a boil-over is an entirely different phenomenon from a slop-over or froth-over. Slop-over involves a minor frothing that occurs when water is sprayed onto the hot surface of a burning oil. Froth-over is not associated with a fire but results when water is present or enters a tank containing hot viscous oil. Upon mixing, the sudden conversion of water to steam causes a portion of the tank contents to overflow.

(8) Water ***Solubility*** of Flammable Liquid

Information of the degree to which a flammable liquid is soluble in water is useful in determining effective extinguishing agents and methods. Alcohol-resistant type foam, for example, is usually recommended for water-soluble flammable liquids. Water-soluble flammable liquids may also be extinguished by dilution, although this method is not commonly used because of the amount of water required to make most liquids nonflammable and there may be danger of frothing in this method if the burning liquid is heated to over 100℃.

3. Classification of Hazards

This section denotes selection of extinguishing agents for the specific class (es) of occupancy hazards to be protected and is classified as follows:

(1) Light (Low) Hazard

Light hazard occupancies are locations where the total amount of Class A combustible materials, including furnishings, decorations, and contents, are of minor quantity. This may include some buildings or rooms occupied as offices, classrooms, assembly halls, and so forth. This classification anticipates that the majority of content items are either noncombustible or so arranged that a fire is not likely to spread rapidly. Small amounts of Class B flammables are included, provided that they are kept in closed containers and safely stored.

(2) Ordinary (Moderate) Hazard

Ordinary hazard occupancies are locations where the total amount of Class A combustibles and Class B flammables are present in greater amounts than expected under light (low) hazard occupancies. These occupancies could consist of offices, classrooms, mercantile shops and allied storage, light manufacturing halls, research operations centers, auto showrooms, parking garages, workshops or support service areas of light (low) hazard occupancies, and warehouses containing Class Ⅰ or Class Ⅱ commodities.

(3) Extra (High) Hazard

Extra hazard occupancies are locations where the total amount of Class A combustibles and Class B flammables present in storage, production, use, and/or finished product is over and above those expected and classed as ordinary (moderate) hazards. These occupancies could consist of wood-working, vehicle repair, aircraft and boat servicing, individual product display showrooms, product convention center displays, and storage and manufacturing processes such as painting, dipping, coating, and including flammable liquid handling refineries, petrochemical gas treating plants, and stations. Also included is warehousing of or in-process storage of other than Class Ⅰ and Class Ⅱ commodities.

New Words and Expressions

vapor [ˈveɪpə (r)]　　n. 蒸汽，烟雾

flammable material　　易燃物品，易燃材料，易燃物

extinguish [ɪksˈtɪŋgwɪʃ]　　v. 扑灭，熄（灯），废除，压制

propagate [ˈprɒpəgeɪt]　　v. 传播，繁殖，宣传，增殖

refer to　　提及，交付

enclosure [ɪnˈkləʊʒə]　　n. 圈地，围场，圈占地，圈用地

flash point　　［油气］［物］闪点，燃点

catalytic [ˌkætəˈlɪtɪk]　　adj. 接触反应的，起催化作用的，催化剂，刺激因素

solubility [ˌsɒljʊˈbɪləti]　　n. 溶解度，可解决性

Exercises

1. What are hazardous locations? What is fire?
2. What are flammable substances and how are they classified?
3. Explain the fire triangle.
4. What is oxidizer? How does an oxidizer work?
5. Give an example of fire control by analyzing a practical case.

科技英语翻译与写作——成分转换

在科技英语翻译时，把原句中的某一成分转译为汉语中的另一种成分的方法就是成分转换法。英语和汉语中都有主语、谓语、宾语、定语、状语和表语，在英语和汉语中的句子结构和表达方式不同，因此，在翻译时会遇到成分转换的情况。

1. 主语转换

（1）转换成宾语

英语的被动语态转译成汉语的主动语态时，可将原文的主语转换成宾语。

Data on the liquid and vapor capacity of equipment of the type being considered for use are needed to determine the necessary cross-section area and size of the equipment.

为了确定设备的必要横截面面积和大小，就需要知道这类设备的液相和气相容积的数据。

If a generator is mot provided, *a battery system* with automatic charging features should be provided.

没有发电机，则应配备有自动充电能力的蓄电池系统。

（2）转换成谓语

Care should be taken at all times to protect the instrument from dust and damp.

应该始终注意保护仪器不沾染灰尘和受潮。

Evaporation *emphasis* is placed on concentrating a solution rather than forming and building crystals.

蒸发着重于将溶液浓缩，而不是生成和析出晶体。

（3）转换成定语

英语句子中的主语转换成汉语中的定语，是为了强调英语句子中的某一种成分，并且把这种成分译成主语。

Medium carbon steel is much stronger than low carbon steel.

中碳钢的强度比低碳钢大得多。（强调“stronger”）

These lathes differ widely in their efficiency after operation.

经过操作，这些车床的效率不大相同。（强调“efficiency”）

（4）转换成状语

The emphasis on efficiency leads to the large, complex operations which are characteristic of engineering.

由于强调效率，因此引起工程所特有的工序繁多而复杂的情况。

2. 谓语转换

（1）转换成主语

A sketch *serves to* express one's idea graphically.

草图的作用就是把人们的想法用图表示出来。

In iron boiler zinc *corrodes* much more rapidly than iron.

在铁质锅炉中，锌的腐蚀要比铁快得多。

（2）转换成定语

The sun *produces in three days* more heat than all the earth fuels could ever produce.

太阳在三天内发出的热量就比地球上所有燃料所发出的热量还要多。

3. 宾语转换

(1) 转换成主语

Semiconductors have a lesser *conduction capacity* than metals.

半导体的导电性能比金属差。

A centrifugal compressor has an apparent *advantage* in that it can be used as a blower or an exhauster.

离心压缩机的明显优点是，它既可用作鼓风机又可用作抽风机。

(2) 转换成谓语

Physical changes do not result in *formation of* new substances, nor do they involve a change in composition.

物理变化不生成新物质，也不改变物质的成分。(省略“result in”)

The modern world is experiencing rapid *development* of science and technology.

当今世界科学技术正在迅速发展。

4. 定语转换

(1) 转换成主语

As has been said before, there are three states *of matter.*

如前所述，物质有三态。

The effects of some chemical substances are *raised blood pressure, inhibited digestion and increased heartbeat.*

某些化学物质会引起血压升高、消化不良和心率过快。

(2) 转换成谓语

Much *strenuous* study of liquid hydrogen as a marine fuel is needed before a determination of its value can be made.

必须大力加强液态氢作为船用材料的研究，才能够确定其价值。

(3) 转换成状语

The theory is of *great* importance that the hotter a body is, the more energy it radiates.

物体越热，其辐射的能量越多，这一理论极为重要。

5. 状语转换

(1) 转换成主语

In this report the aerodynamic characteristics of the new projectile are summarized.

本报告概述了这种新型弹丸的空气动力特性。

When water freezes, it becomes larger in *volume.*

水结冰时，其体积变大了。

(2) 转换成定语

Through the world, coal consumption is growing rapidly.

全世界的煤炭消耗量正在迅速增长。

The critical temperature is different for *different kinds of steel.*

不同类型钢的临界温度是不同的。

(3) 转换成补语

Gas carburization is getting *more generally* used.

气体渗碳法使用得越来越广泛。

The number of molecules in any object is *unimaginably* large.

任何一种物体中的分子数量是大得难以想象的。

6. 表语转换

（1）转换成主语

Wood is not as *dense* as water and therefore it floats.

材料的密度比水小，因而能浮在水面上。

The shortest distance between raw and a finished part is *casting*.

铸造是把原材料加工成成品的最简便的方法。

（2）转换成谓语

We are on the *supposition* that nothing in the world moves faster than light.

我们认为，世界上没有一种物质运动速度能超过光速。

Electronics is *the study of the flow of electrons* and the application of such knowledge to practical problems in communications and controls.

电子学研究电子的流动并把这种知识运用于通信和控制的实际问题。

Recent Developments and Practices to Control Fire in Underground Coal Mines

Coal mine fires cause serious threat to the property and human lives. Out break of fire may be dealt with advanced fire suppression techniques like ***infusion*** of inert gases or liquid nitrogen, dynamic balancing of pressure, reversal of underground mine ventilation, application of nitrogen foam, inertisation of goaf, water mist etc.

Mining is considered as a most hazardous and dangerous of peacetime activities. An outbreak of fire in the underground workings of a mine poses a direct threat from the fire itself. Further, an invisible and immediate threat from carbon monoxide poisoning and an explosion, particularly in gassy coal mines is also there. It affects to both persons working underground at the time of the outbreak and to those involved in the subsequent rescue and fire fighting. It ***hampers*** the coal production and sometimes loss of coal winning machinery.

Fires in coal mines may be categorized into two groups viz. , fires resulting from spontaneous combustion of coal; open fires, which are accidental in nature, caused as a result of ignition of combustible materials.

In coal mines, fires are generally caused due to several reasons viz. , sluggish ventilation, high pressure difference, across intake and return airways, loose and fallen coal in the goaf area, electricity, mechanical friction, blasting, welding, explosions and illicit ***distillation*** of liquor.

An uncontrolled fire in an underground coal mine frequently can only be attacked by sealing off fire zone or the entire mine in the worse situations, or flooding the entire mine with water. The intent of sealing is to cut off the oxygen supply and allow the fire to consume available oxygen inside the sealed off area to cease the combustion process.

1. Injection of Inert Gases

The fire triangle consists of three components viz., fuel, oxygen and source of ignition. If one of the components can be removed from the triangle it is impossible to ignite or sustain any fire. Removal of oxygen from air in underground is impossible but its percentage can be lowered down by infusion of inert gases. Based on this concept the use of inert gases to control underground mine fires have long been practiced in India as well as abroad.

Inert gases can be used in fighting mine fires in the following ways:

1) Reducing the oxygen concentration in the air around the seat of fire (to about 12.5%) so that combustion is inhibited.

2) The prevention of gas explosions by introducing sufficient inert gas into the area to dilute the gas composition out of the so called "explosive zone".

3) Reducing the intensity and spread of secondary combustion and cool the area surrounding the fire zone.

There are three types of inert gases that have generally been used to fight mine fires. They are carbon dioxide, combustion gases, and nitrogen.

Among these nitrogen is widely used in Indian mines. Nitrogen is advantageous because of the fact that it is present in air in high percentages and it can be separated by various techniques while the other two gases do not have such ***facility***. The application of nitrogen in various forms to control underground fire has been discussed below.

2. Dynamic Balancing of Pressure

In majority of the underground fires, the affected areas are sealed by explosion proof/isolation stoppings to exclude leakage of air so that inflammation is put out due to lack of oxygen. However, experience has shown that it is often impossible to make these stoppings airtight particularly under considerable pressure difference. Leakage of air can be totally precluded if pressure across stoppings can be neutralized. Pressure difference across the sealed off area/stoppings can be neutralized by a technique called pressure balancing. In this technique balancing of pressure is achieved by judicious adjustment of air flow rate, first through the different branches of the ventilation network around the affected zone and secondly, the remaining pressure is balanced by adjustment of air flow rates through pipes and pressure chambers specially designed for this purpose. The flow through the pressure balancing chambers and connecting pipes is maintained by the ventilation pressure of the mine itself.

Pressure chamber for the purpose of pressure balancing is generally constructed by building a thin brick stopping provided with a small door at a distance of 2–3 m from the isolation stopping. The isolation stopping should be near to the seat of the fire. Two pipes are laid, connecting the pressure chamber to main intake as well as return airways. Air sampling pipe through the isolation stopping is used to measure the pressure differential across the isolation stopping by a manometer. Difference in pressure across the stopping is balanced by adjusting airflow rates through these pipes.

For successful application of the method of dynamic balancing of pressure following conditions need to be satisfied:

1) Sealed area should not have connection to surface through cracks, fissures etc.

2) The amount of pressure difference that can be neutralized by adjustment of air flow through ventilation circuits is dependent on the ventilation layout, location of the affected zone. The extent of air flow adjustment is possible through ventilation circuits without adversely affecting the ventilation of the mine.

To neutralise the pressure differential across a fire stopping the following procedure are to be adopted.

1) Pressure drop across the fire stopping should be monitored carefully for 24 h to know the range of pressure variation.

2) Air flow rates through relevant circuits of the mine are then adjusted to an extent possible (without unduly affecting ventilation of any working district) and in a manner such that pressure differential across the fire stopping is reduced to minimum possible value.

3) The remnant pressure across the fire stopping and that caused by diurnal variation of atmospheric pressure or that produced by any change in ventilation system of the mine is to be neutralized by adjustment of airflow rate through the pressure chamber made for that purpose and pipes connecting the pressure chamber to main intake as well as return airway.

The proposed device is capable of balancing pressure by automatic adjustment of airflow to pressure chamber by controlling valves in the pipes through a microprocessor based system.

The microprocessor based automatic pressure balancing system consists of three parts:

1) Differential pressure sensor.

2) A microprocessor programmed to receive and process the signal from the differential pressure sensor and then activate airflow control system.

3) Airflow control system comprising of two pipes fitted with stepper motor controlled valves that are guided by signals from the microprocessor.

3. High Pressure Foam

Use of foam plugs has been successful in fighting mine fires in roadways where direct attack with water is not possible. USBM studies reveal that the water content of the foam should not be less than 0. 20 kg/m^3 otherwise the foam is not capable of controlling the fire. With sufficient ventilating forces properly generated foam may be transported over 300 m. Foam does not appear to be effective against deep seated, rapidly advancing, buried or dead end fires.

In India, suppression of spontaneous heating by high pressure high stability foam is a new and active (effective) method. However, the method has been widely used in Czech mines in controlling spontaneous heating of the mined out areas of long-wall panels.

The foam is produced by high pressure foam generator under the pressure of foaming gas. The produced foam is transported by pipelines or fire hoses to the fire area. Inert gas (N_2, CO_2), compressed air or a combination of both is used as foaming gas. The foam generator consists of two independent units namely pumping unit and foam generating unit. The foam is produced from a mixture consisting of water and 5% foaming agent. This mixture is pumped by a pumping unit into a foam-generating unit where the foam is produced. At the same time inert gas (N_2) is supplied to the foam-generating unit at a minimum pressure of 0. 2 MPa, mixed with foaming mixture sprayed from nozzles and then passes through a fine mesh installed inside the foam generation unit. At the outlet of the foaming unit a fire resistant hosepipe of suitable diameter is attached by which the foam is transported to the place of infusion.

4. Water Mist

Water can be used in mines either in the form of spray or mist. McPherson mentioned that once a fire has been progressed to a fuel rich condition there is little chance of extinguishing without sealing off the fire. He does, however, suggest that a means available to gain control of the fire by the application of water as a natural scale fog.

"Water mist" refers to fine water sprays in which 99% of the volume of the spray is in droplets with diameter less than 1,000 ul. Water Mist Fire Suppression Systems (WMFSSs) are readily available, simple in design and construction, easy to maintain, actual in suppressing various fires, non-toxic, and cheaper than other familiar fire suppressing system with no harmful environmental impact. While applied in fire areas, it cleans the air by dissolving soluble ***toxic*** gases produced during combustion, washing down smoke and suppressing dust, and thus improves visibility as well. Unlike many other fire fighting systems, WMFSSs can be safely used in manned areas and found to be active in open condition. Furthermore, water consumption in WMFSSs is ***far less*** than that in water flushing, spraying or sprinkling systems. ***On account of*** these advantages, much study has been carried out in recent years to develop appropriate WMFSSs to control various types and size of fires.

A survey carried out in 1996 indicated that nearly 50 agencies around the world were involved in the research and development of WMFSSs, ranging from theoretical investigations into extinguishing mechanisms and computer modeling to the development, patenting and manufacturing of water mist generating equipment. Water mist is being evaluated for the suppression of fires in diesel fuel storage areas in underground mines at National Institute for Occupational Safety and Health (NIOSH), Pittsburgh. Water mist has shown a positive impact to control a fuel-rich duct fire when a series of experiments on water mist was carried out in a 30 cm square, 9 m long wind tunnel constructed in the Department of Mining & Minerals Engineering, Virginia Polytechnic Institute & State University. A fire is called fuel-rich when the oxygen concentration falls to below 15% in products of combustion.

The concept of water mist to suppress the mine fire is a unique one and for the first time in India it has been tried in the Mine Fire Model Gallery to work out the strategy to control fire with the water mist in actual mining condition. For the purpose an ***indigenous*** system for generation of water mist has been developed.

New Words and Expressions

infusion [ɪnˈfjuːʒ(ə)n]　　n. 注入，灌输，沏成的饮料

hamper [ˈhæmpə]　　v. 妨碍，束缚，限制

distillation [ˌdɪstɪˈleɪʃn]　　n. 蒸馏法，蒸馏，蒸馏作用，蒸馏物

inert gas　　惰性气体

facility [fəˈsɪləti]　　n. 设施，设备，天赋

toxic [ˈtɒksɪk]　　n. 毒物，毒剂

far less　　远不及，大大小于，少得多

on account of　　由于，基于

indigenous [ɪnˈdɪdʒɪnəs]　　adj. 土著的，原产的，本乡本土的，土生土长的

Unit Nine

Gas Outbursts in Coal Mines

Violent ejections of coal and gas from the working coal seam have plagued underground mining operations for over a century. These phenomena are referred to as instantaneous ***outbursts*** and have occurred in virtually all the major coal producing countries and have been the cause of major disasters in the world mining industry. The coal and gas outbursts range in size from a few tonnes to thousands of tonnes of coal with corresponding gas volumes from tens of cubic metres to ***hundreds of thousands*** of cubic metres. In fact, coal and gas outbursts can release over one million cubic feet of gas, fractured and even pulverized coal and rock. Thus, the occurrence of coal and gas outbursts in coal mines and caverns has posed great potential threat to facility operators and has challenged researchers from the rock mechanics and rock engineering community. In the 20th century and a half, since the first reported coal and gas outburst occurred in the Issac Colliery, Loire Coal Field, France, in 1843 (Lama and Bodziony, 1998), it is estimated that as many as 30,000 outbursts have occurred in the world coal mining industry.

These disastrous mine outbursts have resulted in loss of equipment, production time, even entire mines, and sometimes the lives of numerous miners all over the world. A large portion of these was due to secondary factors, such as the following gas explosion, ***suffocation***, and poisoning. Similar disasters have be fallen mines in many other countries. These occurrences have forced mining leaders and researchers to develop an understanding of the complex phenomenon, and procedures to minimise the effect of outbursts or to eliminate them completely. Some safety procedures that have been adopted do lead to reduced production rates.

For almost half a century now, considerable research attention has been paid to this complex problem (Beamish and Crosdale, 1998). The preliminary investigations concerning the coal and gas outburst mechanism through in situ observation, physical and theoretical studies, and numerical modelling, have been made to mitigate the coal and gas outburst hazard during the last decades. Some empirical hypotheses, criteria and analytical models have been proposed for understanding, analysis and prediction of the coal and gas outburst conditions. Kidybinski (1980) took into consideration the three components of gas content and flow, stress, and coal failure and proposed the presence of three zones in the coal seam ahead of the mining operations starting at the coal face: ① protection/degassed zone, ② high gas pressure/active zone and ③ ***abutment*** pressure zone.

Within this model, three fundamental conditions are assumed to be met for an outburst to occur: ① failure of the coal in ***compression*** within the active zone; ② ***penetration*** of a hole through the protection zone; and ③ fluidised bed outflow of the products from the outburst. Gray considered two gas-initiated-failure mechanisms to exist: either tensile failure of unconfined coal or piping of sheared material. Paterson took the general view that, when gas is released from coal, there are body forces on the coal equal to the pressure gradients of the flowing gas. His models therefore were based on the fundamental assumption that an outburst is the structural failure of coal due to excess stress resulting from these body forces. A model proposed by Litwiniszyn was based on the gas existing in a condensed state within the coal. When a shock wave passes through the coal, a phase transformation occurs of the substance into a gaseous state. This sudden creation of gas causes the skeleton of the medium to be destroyed and an outburst is initiated. Support for this model is found in the following observations: ① sometimes "bumps" and instantaneous outbursts occur together, and some "bumps" are regarded as initiation of instantaneous outbursts, and ② in hand-working, especially without the noise of machinery, successive "knocks" in the coal were often precursors to an instantaneous outburst. However, Paterson identified several flaws in this model, in particular cause and effect; where do the shock waves originate? Thermodynamic descriptions have also been proposed for outburst modelling. Williams and Weissmann used a schematic of an outburst in frequently encountered Australian conditions to discuss gas content thresholds for outbursts. They placed emphasis on a gas pressure gradient existing ahead of the working face. However, they also believed "the most important parameter is gas desorption rate, in conjunction with the gas pressure ***gradient*** ahead of the face" (see Figure 9-1).

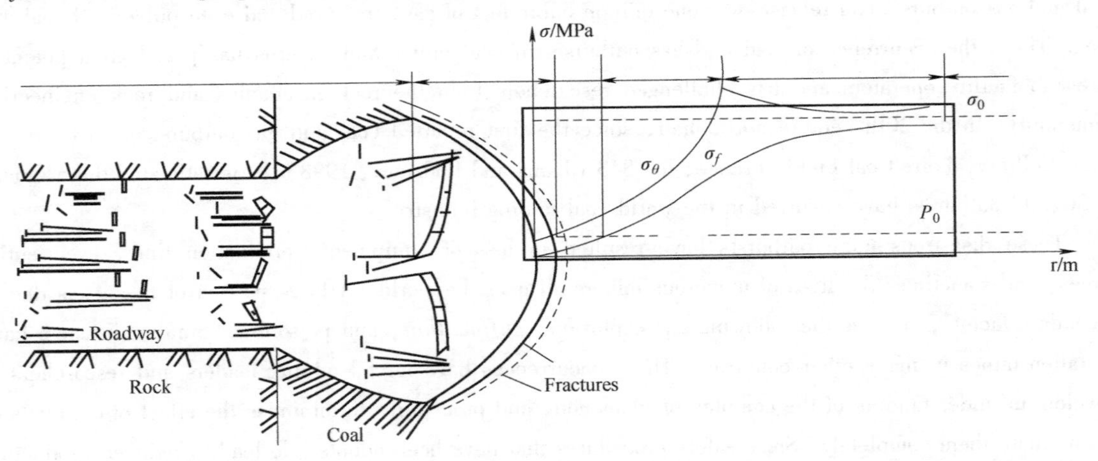

Figure 9-1 Spherical Shell Losing Stability Model during Outbursts (after Jiang and Yu, 1998)

Jiang and Yu, based on many laboratory tests, presented the spherical shell losing stability model during outbursts. They believed that the outburst process consists of six phases, viz., ① intact stress phase, ② stress concentration phase, or abutment pressure phase, ③ coal crushed by rock stress, ④ coal split by gas pressure, ⑤ expulsion of coal and gas due to spherical shell losing stability, and ⑥ movement of coal and gas ***desorption***.

In addition, many previous studies have considered tectonic deformation and the micro-structure of the deformed coal to be important factors influencing outburst occurrence. Farmer and Pooley (1967) found that outbursts only occur in districts subject to severe tectonic movement—hence, their association in many places with ***anthracite***—and in association with such deformational and depositional structures as folds, faults, rolls and slips and, in particular, with rapid fluctuations in the seam thickness. Shepherd et al. (1981) reported on outburst occurrences in Australia, North America, Europe, and Asia, and found that probably over 90% of significant outbursts have been concentrated in the narrow strongly deformed zones along the axes of structures such as asymmetrical ***anticlines***, the hinge zones of recumbent folds, and the intensely deformed zones of strike-slip, reverse, and normal faults. These narrow deformed zones, whether in mesoscopic or mine-scale geological structures form the loci for stress and gas concentrations.

Similar studies in China revealed that outbursts have nearly always occurred in long, narrow outburst zones along the intensely deformed zones of strike-slip, reverse or normal faults, within which coal has been physically altered into cataclastic, granular, or mylonitic micro-structures. The other occurrences are associated with bedding-plane faults and intense folds, which may produce these micro-structures in broader zones. In either case, the outburst-prone zones generally cover no more than 20% to 30% of the mine area. Some fault zones do not exhibit altered micro-structure, and these are not prone to outburst. Thus, the presence of these altered micro-structures is considered as the first essential factor for outburst occurrence, and outburst-prone districts could be predicted by studying the spatial distribution of altered coal and geological structures. It has also been found that outburst danger increases with the intensity of deformation and alteration of the coal micro-structure. Many studies have compared coal actually expelled from an outburst cavity to coal in situ with similar micro-structure, based on their physical and morphological characteristics. To date no significant difference has been found.

Naturally, a deeper understanding of the outburst mechanism and reliable methods for the prediction of outbursts must be not only based upon long years of practical experience in mines, but also on scientific research and experimentation. Despite extensive research about the violent coal and gas outbursts which have occurred in coal mines, surprisingly little progress has been achieved in the past 150 years towards understanding or prediction. In particular, a quantitative model that describes the progressive failure process, as well as the violent outbursts process in coal mines, has not appeared. It is the aim of this passage to present such a model and to show how the model explains the observations associated with outbursts.

The numerical model shown in Figure 9-2 is designed to simulate the instantaneous outbursts which occurred in the course of cross-cutting induced by mining. In the model, the gassy soft coal seam is enclosed by an impermeable hard roof and floor rock. Moreover, a layer of thick hard rock acts as a protective screen ahead of the coal seam. The layer of hard rock is instantaneously opened by drilling and the coal seam behind the protective screen is therefore exposed in the course of this cross-cutting. The simulation model is discretised into a 150×200 mesh (30,000 elements). The gas pressure saturated in the coal seam is 2.1 MPa and the Young's modulus and compressive strength of the coal are 10 GPa and 15 MPa, respectively. In addition, the Young's modulus and strengths of the rock roof and floor far exceed those of the coal seam.

The mechanical and seepage parameters in the numerical model are presented in Table 9-1.

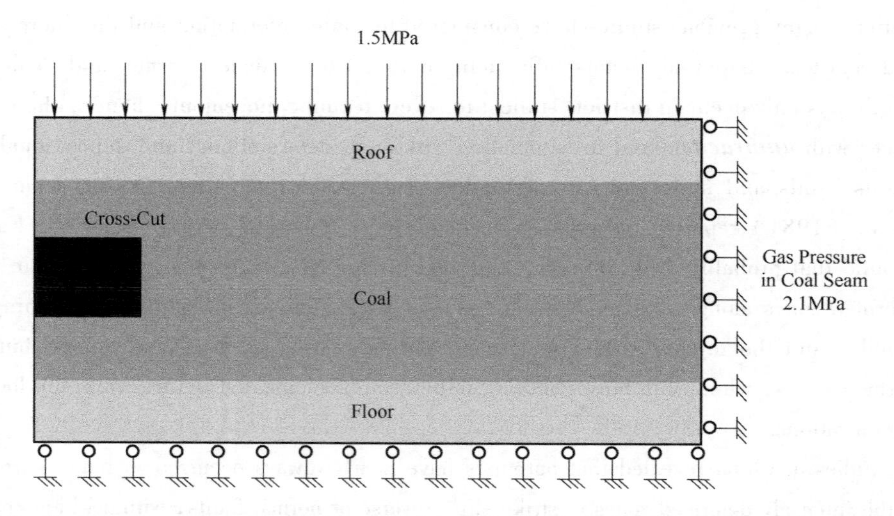

Figure 9-2　Numerical and Seepage Model of Coal and Outbursts Induced by Numerical and Spenetration

Table 9-1　Mechanical and Seepage Parameters Used in the Numerical Model

Mechanical and Seepage Parameters	Cool Seam	Roof and Floor	Cross-Cut
Homogeneity index	2	10	2
Mean elastic modulus, E_0/GPa	5	50	10
Mean compressive strength, σ_0/MPa	100	300	150
Internal friction angle, Φ/ (°)	30	32	30
Bulk weight/ (kg/m^3)	2. 7	1. 4	2. 0
Ratio of UCS to UTS	20	10	10
Poisson's ratio, ν	0. 3	0. 25	0. 3
Gas permeability, $\lambda/[m^2/(MPa^2 \cdot d)]$	0. 1	0. 001	0. 01
Coefficient of gas content, A	2	0. 1	0. 1
Coefficient of pore pressure	0. 5	0. 01	0. 1

It can be seen from the ***numerically simulated*** results induced by cross-cutting that the whole process of coal and gas outbursts can be divided into four stages:

1) Stress concentration stage. At the beginning of cross-cutting, the loads from the upper rock strata are mostly carried on the freshly exposed coal due to stress concentrations. The stress in the coal is not uniformly distributed at the mesoscopic scale because of the heterogeneity of the materials.

2) Coal/rock fracture and splitting induced by rock stress. Micro-fractures in coals are predominant under the abutment stress in this stage. The mechanical properties of coal progressively degrade due to the effect of the stress concentration and creep, as well as the three-dimensional stress state in the coal near the working face gradually transforming into a two-dimensional stress state. As a result, fracturing and splitting parallel to the free exposed face first occurs in coal near the working face. In the course of splitting, the stress in the coal near the coal face decreases and the peak stress location gradually moves away from the coalface and into the deep coal, with release of the elastic energy stored in the coal. It is noticeable that a cluster of cracks begins to form along with the transfer of the stress peak.

3) Crack propagation driven by gas pressure. High-pressurised gas saturated in the coal seam gushes into the "gas way" and quickly and violently splits the fractured coal. The effect of high gas pressure

eventually leads to the formation of the "gas way" during the propagation and coalescence of clusters of cracks.

4) Ejection of coal induced by gas pressure, i. e. outbursts. During the process of crack propagation and coalescence induced by gas pressure, the cracks volumetrically expand and the gas gushes into the cracks. There is a great gas pressure gradient because the gas pressure saturated in the cracks is far beyond the pressure in the freshly exposed coalface, which causes the ejection of the fractured and splitting coal, i. e. coal and gas outbursts.

The numerical simulated results reveal that in situ stress, gas pressure and the physico-mechanical properties of coal and rock are the main contributing factors affecting coal and gas outbursts. In addition, numerical simulated results not only trace the initiation, propagation and coalescence of cracks in the coal, but also present the associated evolution of the stress field in the coal seam and the roof and floor of the rock strata, i. e., the stress redistribution in the coal seam and rock roof and floor at every stage.

New Words and Expressions

outburst ['aʊtbɜːst]	*n.* （蒸汽，怒气等的）爆发，突发
hundreds of thousands	成千上万
suffocation [ˌsʌfə'keɪʃən]	*n.* 窒息，闷死
abutment [ə'bʌtm(ə)nt]	*n.* [建] 桥台，邻接，接合（点）
compression [kəm'preʃ(ə)n]	*n.* 压缩，加压，压紧，浓缩
penetration [penɪ'treɪʃ(ə)n]	*n.* 渗透，突破，侵入，洞察力
gradient ['greɪdɪənt]	*n.* 梯度，坡度，倾斜度
desorption [dɪ'sɔːpʃən]	*n.* 解吸，释放，[化] 脱附，脱附作用
anthracite ['ænθrəsaɪt]	*n.* [矿物] 无烟煤
anticline ['æntɪklaɪn]	*n.* [地质] 背斜，背斜层
numerically simulated	进行数值模拟，数值模拟

Exercises

1. What are gas outbursts in coal mines?

2. Why has the occurrence of coal and gas outbursts in coal mines and caverns posed great potential threat to facility operators and has challenged researchers from the rock mechanics and rock engineering community?

3. What are the three fundamental conditions considered in Kidybinski's model?

4. Explain the spherical shell losing stability model presented by Jiang and Yu?

5. Explain the numerical and seepage model of coal and gas outbursts introduced by "cross cutting" penetration.

科技英语翻译技巧——定语从句

由于科技英语文献的强概念性和强逻辑性，定语从句出现的频率很高。定语从句可修饰名词、代词或者名词性词组，可对所要说明的词语进行补充说明，使语句更加完整、清楚。同时，定语从句还可用于补充、转折、因果、目的、条件、让步等。定语从句的翻译方法可分为合译法和分译法两种。

1. 合译法

合译法是指在翻译时，把从句混合到主句中，放在所修饰名词之前，译成汉语的单句。

（1）定语译法

句子的结构较为简单，且在逻辑上有明确的先行词的定语从句，若翻译时将从句置于先行词之前不影响表达时，可合译。例如：

Many of the same engineers and companies that had established formal system safety defense programs were also involved in space programs, and the systems engineering and system safety technology and management activities were transferred to this new work.

许多建立了正式系统安全防御项目的工程师和公司都参与了航天项目，在这个新的项目中，采用了系统工程、系统安全技术和管理活动。

New hazard analysis approaches that include software were used.

新的危险分析方法还包括软件的使用。

This case study will describe these three investigations in terms of the aspect of ergonomics that was applied and the effect of that application on the outcome of the investigation.

这个案例将描述这些应用在人机工程学方面的三项调查和调查结论运用的效果。

（2）译成谓语

当定语从句是全句的重点时，应将从句顺序译成简单句中的谓语，把主句压缩成主语。例如：

Another explanation is that the changes brought about by the mine's redistribution of stress trigger latent seismic events, deriving from the strain energy produced by its geological aspects.

另一种解释是由于矿山压力重新分配带来的变化引发了潜在的地震，它们源于地质方面产生的应变能。

2. 分译法

分译法就是将定语从句与主句分开来译。分译时可根据不同的情况，译成并列句或者状语从句。

（1）译成并列句

Fault Tree Analysis (FTA) is a failure analysis in which an undesired state of a system is analyzed using boolean logic to combine a series of lower-level events.

事故树分析（FTA）就是失效分析，即用布尔逻辑结合较低级别的一系列事件，对系统的不理想状态进行分析。

The manufacturer or supplier has a duty to provide an instruction manual which explains the erection sequence, including any bracing requirements.

制造商或供应商有责任提供指导手册，说明安装顺序，包括所有支撑的要求。

The opening of mine shaft relieves neighboring rocks of tremendous pressure, which can literally cause the rock to explode as it attempts to re-establish equilibrium.

矿井裂缝会缓解周边岩石的巨大压力，这实际上会引发岩石的爆炸，因为它试图重新建立平衡。

（2）译成状语从句

当定语从句对其先行词没有明显的修饰和限制作用，却有原因、结果、目的、条件、时间、让步状语等意义时，可翻译成状语从句。

Process-engagement risk may be an issue when ineffective operational procedures are applied.

当应用无效的作业程序时，参与风险可能会成为一个问题。

The first legislation for mine dust appears to have been formulated in 1912 when the Union of South Africa introduced laws governing working conditions in the gold mines of the Witwatersrand.

针对矿尘的第一次立法好像在1912年就已经制定，当时南非联盟已出台一些法规来监督在斯兰德金矿的工作条件。

The overall requirement is that all persons must be able to work and travel within an environment that is safe and which provides reasonable comfort.

总的要求是，所有的人必须能够在安全、舒适的环境下工作和旅行。

Greenhouse Gas Emissions from Australian Open-Cut Coal Mines Contribution from Spontaneous Combustion and Low-Temperature Oxidation

In Australia, coal mining is a significant source of greenhouse gases. During 2006, emissions from the coal mining industry (both underground and open-cut) were estimated to contribute around 6% of the nation's total emissions (Australian Dept of Climate Change 2008a). Slightly more than half of these emissions were attributed to fugitive emissions of methane, which is released as the coal is mined, while most of the remainder resulted from energy usage.

Open-cut mining also produces large quantities of ***carbonaceous*** waste material, which, when exposed to air, is subject to low-temperature ***oxidation*** and, in some cases, spontaneous combustion. These processes produce CO_2 and other gases, including methane, and therefore represent a potential additional source of greenhouse emissions. Worldwide, the significance of emissions from coal fires is well known (Stracher and Taylor 2004; Kuenzer et al. 2007), but quantitative data on the scale of these emissions are rare. In a study of greenhouse emissions from the Australian coal industry, Williams et al. suggested that up to 25% of the total greenhouse emissions from an open-cut mine may be derived from spontaneous combustion and low-temperature oxidation; however, this estimate was based on very limited data.

The Intergovernmental Panel on Climate Change recognises the potential of low temperature oxidation and spontaneous combustion as a source of greenhouse gases, but at present, there is no recognised method for making reliable estimates. As a consequence, emissions from this source are not presently included in the Australian National Greenhouse Gas Inventory. However, with the likely introduction of an emissions trading ***scheme*** in Australia, it is becoming increasingly important to be able to ***account for*** all sources of emissions.

Apart from the Williams et al. report, we are unaware of any other published estimates of greenhouse emissions from spontaneous combustion and low-temperature oxidation at the individual mine level. Hence, the scale of these emissions is largely unknown. Accordingly, research into developing a tractable methodology for estimating emissions has been conducted in Australia over the past decade. Carras et al., for instance, recently reported results of a study conducted at 11 Australian open-cut coal mines where emissions of CO_2

and CH_4 from low-temperature oxidation and spontaneous combustion in spoil and other waste were measured using flux chambers. In that study, emissions were categorised according to the ***intensity*** of spontaneous combustion, and despite considerable ***scatter*** in the data, three broad categories of spontaneous combustion were identified. Each category was assigned an emission factor ranging from about 12 to 8200 kg CO_2-e $year^{-1}m^{-2}$. With appropriate classification of the affected ground, these emission factors provide the basis for estimating emissions from spontaneous combustion and low-temperature oxidation of coal wastes.

Other work has investigated the use of airborne infrared thermography to measure the surface temperature of spoil piles. This, combined with the finding that the emission flux is roughly proportional to the surface temperature, also provides a potential method for estimating emissions over relatively large areas. To date, however, these methods have not been applied to quantify emissions from individual mines.

In the study reported in this paper, estimates of emissions from spontaneous combustion and low-temperature oxidation of waste material were made for a number of operating open-cut coal mines in Australia using the methods developed by Carras et al. To place these estimates into context, emission data from all other sources at each mine were also considered and compared with the estimates made for spontaneous combustion and low-temperature oxidation.

Greenhouse gas emission data were analysed from six open-cut coal mines located in two of Australia's major coal producing regions: four in the Hunter Valley in New South Wales (NSW) and two in the Bowen Basin in Queensland. Annual production of run-of-mine (ROM) coal from these mines ranged from approximately 1.7 Mt to more than 16 Mt. Mines were selected to include those with low, ***medium*** and severe occurrences of spontaneous combustion.

Data on electricity, fuel and explosives consumption, land clearing, ***fugitive*** emissions and other sources of greenhouse gas sources were obtained from each mine. Information on the extent and severity of spontaneous combustion was also collected. For two of the mines selected, infrared data from the Carras et al. study were available, which were also used to estimate spontaneous combustion derived emissions from these mines.

1. Fugitive Emissions from Seam Gas

Fugitive emissions of methane and, to a lesser extent, CO_2 represent a large proportion of greenhouse gas emissions from the Australian coal mining industry. For ***open-cut mining***, however, these emissions are not readily measured. Saghafi et al. recently proposed a method for estimating fugitive emissions from individual mines; however, at present, there are insufficient data available for most Australian mines to allow reliable estimates to be made. Consequently, estimates of fugitive emissions were made by multiplying the annual ROM coal production by an emission factor. This is the method currently used for compiling the Australian National Greenhouse Gas Inventory.

2. Emissions from Spontaneous Combustion and Low-Temperature Oxidation

Emissions from spontaneous combustion were initially estimated by determining the area of ground affected by spontaneous combustion and applying an emission factor based on the work of Carras et al. (2009). Those authors observed a correlation between the surface temperature of a spoil pile and surface emission rate. While acknowledging that the surface temperature can at best be a crude measure of spontaneous combustion activity within a spoil pile and that the emissions are also governed by the nature of the topmost layers of spoil, the authors nevertheless found a useful relationship. This is shown in Figure 9-3 (from Carras et al. 2009).

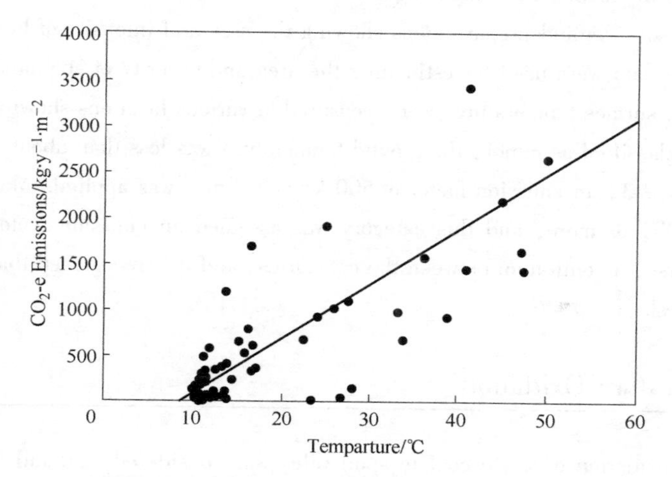

Figure 9-3 Average Emission Rates from Spoil Affected by Spontaneous Combustion (as CO_2-equivalent) as a Function of Average Surface Temperature (from Carras et al. 2009)

Methane is often formed and emitted during spontaneous combustion, and the emission factors developed by Carras et al. (2009) took this into account.

Linear regression of the data presented in Figure 9-3 yielded the empirical expression shown in Equation 9-1, where the emission rate, ER, in units of kg CO_2-e year^{-1} m^{-2} is related to the surface temperature, T (℃), by the expression:

$$ER = 59.6T - 496.4 \tag{9-1}$$

Equation 9-1 is based on measurements made when the ***ambient*** temperature of unaffected ground was about 9℃ (i. e., threshold temperature below which there are no emissions). Applying Equation 9-1 to datasets measured when the threshold temperature is significantly different, such as during summer months, may substantially over-or underestimate emissions. Hence, care was taken to ensure that ground temperature measurements were made in the early morning during winter months.

Estimates of the area of ground (in square meter) affected by spontaneous combustion in each mine were made by visual inspection and the intensity categorised as either "minor" or "major" according to the following criteria:

- minor—active spontaneous combustion with less obvious signs;
- surface discolouration;
- no large cracks;
- little smoke or steam;
- major—active spontaneous combustion with marked surface signs;
- large cracks with obvious signs of gas venting;
- smoke and steam;

- hot gases;
- surface discolouration.

This information was gathered during site visits to mines and detailed discussions with mine personnel. Because of statutory environmental licensing ***obligations***, a number of the mines already collect this information through regular ground surveys and prepare maps showing the area and intensity of heating within the mine. Where available, these data were used for estimating the area and severity of spontaneous combustion.

During site visits, surface temperatures were measured at various locations showing signs of spontaneous combustion. In areas classified as minor, the ground temperature was less than about 10℃ above unaffected ground so, from Figure 9-3, an emission factor of 500 kg $year^{-1}m^{-2}$ was applied. Major outbreaks typically had temperatures of 60℃ or more, and this category was assigned an emission factor of 3,000 kg $year^{-1}$ m^{-2}. There is, of course, a continuum between the categories, and it is recognised that this approach is only an approximation at best.

3. Low-Temperature Oxidation

Low-temperature oxidation of waste coal in spoil piles was considered in detail by Carras et al. Using the results of laboratory measurements of oxidation rates of coal and carbonaceous material as well as numerical modelling, they calculated greenhouse gas emission rates from mining waste for a range of carbon contents and temperatures. They found that the emission rates from a variety of sources, including bare spoil, revegetated spoil, uncovered tailings, uncovered reject, natural forest floor and suburban ***lawn***, were all similar. From this, they suggested that greenhouse emissions of CO_2 from rehabilitated spoil, with no spontaneous combustion, could not be distinguished from other natural land surfaces, although they also indicated that further work would be required to ***disentangle*** the biogenic and coal-based contributions.

For the purpose of the current study, we assumed that the average temperature and carbon content of the spoil were 27℃ and 5%, respectively. According to the results of Carras et al. (2009), this yields an emission factor of 2.2 kg $year^{-1}m^{-2}$, which was applied to low-temperature oxidation from uncovered spoil, tailings or reject with no spontaneous combustion.

To estimate the contribution from low-temperature oxidation, it is also necessary to know the surface area of exposed spoil. Aerial photographs of mines A, B, C, and D indicated that the spoil areas were roughly 4, 4, 0.5, and 2 km^2, respectively. Photographs of mines E and F were not available so the exposed spoil area was assumed to be 3 km^2 in each case since they were about three quarters the size of mines A and B.

New Words and Expressions

carbonaceous [ˌkɑːbəˈneɪʃəs]   *adj.* 碳的
oxidation [ɒksɪˈdeɪʃ(ə)n]   *n.* 氧化
scheme [skiːm]   *n.* 方案，计划，体制，体系
account for   说明（原因、理由等），导致，引起，占……比例，对……负责
intensity [ɪnˈtensɪti]   *n.* 强度，烈度，强烈，紧张
scatter [ˈskætə]   *n.* 分散，散播

medium [ˈmiːdɪəm]	*adj.* 中等的，中号的
fugitive [ˈfjuːdʒɪtɪv]	*adj.* 逃亡的，逃跑的，短暂的，易逝的
open-cut mining	露天采矿
ambient [ˈæmbɪənt]	*adj.* 周围环境的，周围的，产生轻松氛围的
obligation [ɒblɪˈgeɪʃ（ə）n]	*n.* 义务，责任，债务，负担
lawn [lɔːn]	*n.* 草坪，草地，上等细棉布
disentangle [dɪsɪnˈtæŋg（ə）l]	*v.* 解开……的结，理顺，使解脱，使脱出

Unit Ten

Risk of Coal Dust Explosion and Its Elimination

The phenomenon of a ***coal dust explosion*** has been known since more than 100 years. Almost 100 years ago the Courrieres disaster, costing 1099 lives took place. Other tragedies followed and continue to occur. We would like to be good prophets and tell that the coal dust explosion hazard is over, but experience and statistics tell us that explosions have occurred and will continue to occur. The question appears—is the lesson of many disasters not learned? Do they result because of lack of knowledge and protection means? What is the cause of the biggest mining hazard? Honest answer for these questions is a simple one—knowledge of coal dust explosion causes, how it occurs, and how to avoid it which knowledge—wise is almost complete. But there are troubles with using this knowledge in everyday mining practice. Investigations of separates cases show that breaking of elementary safety rules always caused the explosion. This was proved in the recent catastrophes in Poland (Cybulski, 2002) and abroad (McPherson, 2001). The general conclusion of such investigation is needed for simple and reliable protection systems. We want to consider what really should be done from the point of view of mining technology to avoid a coal dust explosion.

The Courrieres disaster appeared to be a milestone because it showed the dominating role of coal dust in the mining explosion—the importance of the presence of the ***firedamp***, even though as an explosion factor, it is not necessary for explosion to occur. To answer the above questions the simple, general model of an explosion of any ***flammable*** substance is very useful. This model relates to all explosion phenomena with combustion, not only in underground mining. Appearance of explosion requires the simultaneous appearance of five factors, forming the so-called explosion pentagon (see Figure 10-1).

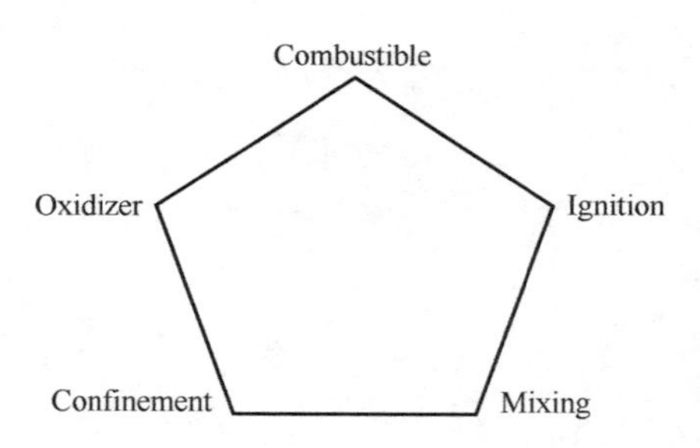

Figure 10-1 Explosion pentagon

Lack of one of the elements of the pentagon or braking connection between them excludes the explosion's appearance. In the case of the underground coal mine this model says that:

- combustible (coal dust) exists by its nature and in the explosion the coal dust participates;
- oxidizer is atmospheric air;
- ***ignition*** sources result from mining technology;
- underground mine workings are by their nature confined rooms where overpressure can grow, without dissipating as it otherwise would in open space; note that there is not an explosion hazard in open cast mining.

Mixing of ***combustible*** and oxidizer or the forming of the dust cloud in air occurs in coincidence with ignition. In a normal state the dust cloud is not in explosion ***concentration*** range for the mine air; the presence of people in such a cloud of dust is not possible.

Those are the main factors of explosion risk in mine working. They enable one to formulate the elementary principles, which upon meeting diminish the explosion risk.

1. What Is the Origin of Dust Explosion Hazard

Among the specialists there is not an unanimous opinion to this question: Is the coal dust explosion a natural hazard or technical one? Dust is created by the technology of ***extraction*** so it is reasonable to treat this hazard as a technical one. But in this case a common understanding of the technical hazard limit is needed by the contact man—this pertains to the machine and a moving object (typical example being falling rock). So, according to Polish regulations dust hazard is considered to be a natural one. As a consequence of such qualification there is the dividing of workings, seams or other part of mine into classes, either categories or other states of hazard. Protection means used in the given working depend on its class.

Openly saying, the current safety regulations, if strictly met, reduce the risk of explosion. Looking at the problem of explosion risk, according to the Standard PN-N-18000 one may say that the risk of explosion in the given working depends on:

- the amount of dust accumulated;
- size distribution of dust;
- depositing of new dust layers on the working surface;
- volatile content of coal;
- incombustible content of layered dust.

Dust accumulated on the floor, sidewalls, pipelines, machines and other devices is not strictly (exclusively) coal dust. It contains limestone dust used for inertization, natural rock dust coming from rock cutting, surface moisture, and other components different than coal. Such a mixture is called "mine dust" or float dust in the US. According to regulations (2002) "mine dust—created during mining and processing of minerals includes substances added to inertize the dust". Nevertheless, in the workings where no inertizing by adding of limestone dust is pursued "mine dust" is very close in character to "coal dust".

In the National Regulations (2002) there are two distinctive notions used: safe dust and unsafe dust. The safe dust contains:

- at least 70% of incombustible if deposited in non gassy fields;
- at least 80% of incombustible if deposited in gassy fields;
- free water in quantity to not enable dust propagation.

Coal dust is created during mining and transported by the ***ventilation*** stream. Workings are covered with dust of different grain size. Cybulski's investigation proved that layered dust actually contains two important components described as being different in size:

- dust "d85" or "fine dust" containing 85% of grain mass passing through a sieve of 75 μm (this dust accumulates at the upper parts of workings);
- dust "d25" or "medium dust" containing 25% passing through the 75 μm sieve (accumulates in lower parts of workings).

Such inhomogeneous distribution of dust causes the phenomenon of "separate dust clouds" —described by Cybulski.

Exact data is lacking on the total amount of dust produced in Polish coal mines. The main sources of dust are:

- shearers and roadheaders;
- roof support movement;
- transportation of minerals particularly at transfer points;
- processing (screening, crushing).

It is admitted that 2%–3% of the extracted rock is transformed into dust, which gives in Polish coal mining an amount of dust up to 3 millions tons a year. It does not mean that all this dust creates a danger of explosion. Although effectiveness of dust suppression systems (water spraying, ***dust collectors*** etc.) is still considered to be insufficient these systems diminish the amount of dust deposited in the roadways. Dust deposition is dynamical; it is a continuous process that increases the amount of dust in the workings. It is characterized by intensity of deposition defined as a mass of dust deposited on the given surface at the determined time. To characterize the explosion hazard only flammable coal dust is taken into account. The measurement unit of the intensity of deposition is gram per square meter per day ($g/m^2/day$).

Amount of deposited dust at a given location depends on:

- intensity of the source expressed in unit of mass of dust created in the unit of time;
- size distribution of dust;
- air flow velocity in the working;
- flow resistance per unit of working's length.

Typical values of intensity of deposition are in range from almost zero to 150 grams per square meter per day. The upper limit did not change since introducing high productive longwalls, with the dusting being bound through the use of modern effective dust suppression means.

Deposited dust is hazardous if raised into air forming a cloud with sufficient concentration to propagate explosion. The ***lower explosion limit*** is determined as 50 grams per cubic meter. Such a big concentration makes a man's breathing impossible: Such a high dust concentration (although being a minimum one) does not normally exist at working places. Being different from flammable gases, entrained dust in typical practice does not reach the upper explosion limit. In mining practice it is very difficult to create a homogenous dust cloud of a concentration more than one kilogram per cubic meter. Although Cybulski determined the upper explosion limit to be one kilogram per cubic meter, it means only that it was the upper estimable concentration formed in an underground gallery. It would be a tragically misunderstanding to consider "upper explosion limit" in the same way as for gases—in the case of mine gases, beyond the upper explosion limit the flame does not propagate.

Amount of dust accumulated in working is determined as nominal concentration—that being measured as the total mass of dust (determined by a special sampling procedure) in given roadway section divided by its volume. The nominal concentrations are in range from grams per cubic meter to kilograms per cubic meter.

2. Approach to Coal Dust Explosion Risk

The European Standardization Committee approved the European Standard EN 1127-2 "Explosive atmospheres: Explosion prevention and protection, Part 2: Basic concepts and methodology for mining". It is a second part of the standard EN 1127-1 "Explosive atmospheres—Explosion prevention and protection: Basic concepts and methodology". Both standards are harmonized with following Directives 98/37/EC, "Machine Safety" and 94/9/EC ATEX Directive. Both standards have been implemented into Polish Law, European Standard EN 1127-2 introduces notions less used in Polish coal industry. Some definitions are:

1) Explosive atmosphere—mixture with air, under atmospheric conditions, of flammable substances in the form of gases, vapours, mists or dust, in which, after ignition has occurred, combustion spreads to the entire unburned mixture.

2) Hybrid mixture—mixture of flammable substances with air in often in a rather complex physical (multi-phase) state. Examples of hybrid mixture: methane-coal dust-air, petrol drops-petrol vapour-air.

3) Potentially explosive atmosphere—atmosphere which could become explosive due to local and operational conditions, and protective system means design units which are intended to halt incipient explosions immediately and/or to limit the effective range of explosion flames and explosion pressures. Protective systems may be integrated into equipment or separately placed on the market for use as autonomous systems.

The notion of explosive atmosphere, being basic in European Standardization in the fields of explosion safety was not previously used in the Polish technical dictionary. Below, is presented the most important statement of the standard; the elements are different from those used in National Regulations:

Hazardous conditions 1 (explosive atmospheres): Underground parts of mine and associated surface installations of such mine endangered by firedamp and/or flammable dust. This includes underworkings where the concentration of firedamp is within the explosion range, e. g. as a result of malfunction (e. g. breakdown of fans), sudden release of large amounts of firedamp (gas outburst) or increased gas emission) due to decrease of air pressure or increase in coal winning).

Hazardous conditions 2 (potentially explosive atmosphere): Underground part of mines and associated surface installations of such mines likely to be endangered by firedamp and/or flammable dust. This includes underworkings where the concentration of firedamp in the ventilating current or in the firedamp ***drainage*** system is outside the explosive range.

ATEX Directive specifies different means of protection, which reflect the requirements of the different atmospheric conditions. Criteria determining the classification into categories are the following.

Category M1 comprises equipment designed and, where necessary, fitted with additional special means of protection to be capable of functioning in conformity with the operational parameters established by the manufacturer and ensuring a very high level of protection.

Equipment in this category is intended for use in underground parts of mines as well as those parts of surface installations of such mines endangered by firedamp and/or flammable dust. Equipment in this category shall remain functional even in the event of rare equipment faults in an explosive atmosphere and have explosion protection measures so that either:

- In the event of failure of one means of protection, at least an independent second means provides the requisite level of protection;
- The requisite level of protection is assured in the event of two faults occurring independently of each other.

Category M2 comprises equipment designed to be capable of functioning in conformity with the operational parameters established by the manufacturer and ensuring a high level of protection.

Equipment in this category is intended for use in underground parts of mines as well as those parts of surface installations of these mines likely to be endangered by firedamp and/or flammable dust.

New Words and Expressions

coal dust explosion 煤尘爆炸，煤粉爆炸
firedamp [ˈfaɪədæmp] *n.* 甲烷，沼气
flammable [ˈflæməb(ə)l] *adj.* 易燃的，可燃的
ignition [ɪɡˈnɪʃ(ə)n] *n.* 点火，着火，点燃，点火装置
combustible [kəmˈbʌstɪb(ə)l] *n.* 可燃物，易燃物，燃料
concentration [kɒns(ə)nˈtreɪʃ(ə)n] *n.* 浓度，含量，集中，专心
extraction [ɪkˈstrækʃ(ə)n] *n.* 提取，开采，提炼，拔出
ventilation [ˌventɪˈleɪʃ(ə)n] *n.* 通风设备；空气流通
dust collector 除尘器
lower explosion limit 爆炸下限
drainage [ˈdreɪnɪdʒ] *n.* 排水，排水系统，放水

Exercises

1. What is coal dust explosion?
2. What is the cause of the biggest mining hazard, how does it occur, and how to avoid it?
3. Explain the general model of an explosion of any flammable substance.
4. What is the origin of dust explosion hazard?
5. What does the amount of deposited dust at a given location depend on?

科技英语翻译技巧——状语从句

英语中的状语从句有多种形式，包括时间状语从句、地点状语从句、原因状语从句、结果状语从句、目的状语从句、条件状语从句和让步状语从句。状语从句可位于主语之前、之后或中间，在翻译时，大多放在主语之前。定语从句的类型较好判断，汉语中有与状语从句的连词相对应的词语，在翻译时，应遵循汉语的表达方式，对状语从句的位置、连词的译法和省略进行调整。

1. 时间状语从句

在翻译时间状语从句时，无论它在原句中的位置如何，都应该按照汉语的习惯，把时间状语从句翻译到主句之前。由 when 引导的时间状语从句，在翻译时可转译成条件状语从句或让步状语从句的形式。

引导时间状语从句常见的连词有：when“当……”，after“在……以后”，before“在……以前”，until或till“直到……才”，as soon as“一……就”，while“正当……时，在……期间”，as“当……时，随着”，once“一旦”。

The conductivity of this material increases as the temperature increases.

该材料的导电率随温度的上升而增加。

Electrical forces, once they have been produced, seem to be very strong.

电力一旦产生后似乎是很大的。

2. 地点状语从句

在翻译地点状语从句时，一般按照从句的位置进行翻译，即从句在主语之前时，翻译时就放在主语之前；从句在主语之后，翻译时就放在主语之后。当然，可根据情况，将主语后面的状语从句翻译在主语之前。当有些地点状语从句有条件状语从句或者结果状语从句的含义时，可按照条件状语从句或者结果状语从句翻译。

引导地点状语从句常见的连词是where“在……地方”。在科技英语中，where还可表示一种条件或时间，意为“在……情况下，当……时”。

It is to provide airflow in sufficient quantity and quality to dilute contaminants to safe concentration in all parts of the facility where personnel are required to work or travel.

在设施内所有人员工作或经过的地方，需要提供足量的、质量优良的气流来稀释污染物浓度到安全值。

When a large enough force is applied to a body, the body will start to move.

当把一个足够大的力加到物体上时，该物体就会开始运动。

3. 原因状语从句

原因状语从句一般翻译在主语之前，但有时考虑到上下文的结构，可翻译在主语之后。

引导原因状语从句的连接词有：because“因为……”，since“因为……，既然……”，as“因为……”，for“由于”，now that“既然，由于”，in that“既然，由于”。

Interim inspections may be needed between detailed inspections because the employer's risk assessment has identified a risk that could result in significant deterioration.

详细检查中间可能需要临时检查，因为经雇主的风险评估确定的风险可能会发生严重恶化。

A transformer cannot be called a machine for it has no moving parts.

变压器不能被称为机器，因为它没有运动的部件。

4. 结果状语从句

翻译结果状语从句时，因汉语和英语的习惯都是把结果放在句末，所以，可采用顺译法。

引导结果状语从句的先行词有：so... that，so that，such... that，enough... that，等。

Yet more and more deadly chemicals are added to the list each year and new uses are devised so that contact with these materials has become practically worldwide.

然而每年却有越来越多新的致命化学物质出现，并且其新的用途层出不穷，因此，与这些物质接触实际上已经全球化了。

The concept of work is so important that it will bear further discussion.

功的概念极为重要，所以还要进一步加以讨论。

5. 目的状语从句

科技英语中的目的状语从句一般采用顺译法，若需要强调目的时，可将从句翻译在主语之前。

目的状语从句常用的引导词有：in order that“为了……”，so that“以便……”，lest“以免”。

Oil spills on water are contained using floating booms and adsorbents, or solid materials that capture the soil, so that it can be disposed of in landfills.

使用浮动栅栏和吸附剂，或使用固体材料捕集土壤，使泄漏到水面上的油得到控制，以便油污能够在填埋区被处理。

The magnet is usually made in the shape of horseshoes so that it will be as strong as possible.

磁铁通常做成马蹄形的，以便使其磁性尽可能地强。

6. 条件状语从句

若条件状语从句在原句中放在了主语之后，那么就可采用顺译法，对主语进行补充说明。而一般条件下，应放在主语之前进行翻译。

条件状语从句常用的引导词有：if “如果”，unless “除非”，so long as “只要”，so far as/as far as “只要，就……而言”，except that “只是，除了……之外”，in the event that “即使，万一”，suppose/supposing “假设”。

If the shot firer needs to remain within the zone, he should be in a safe position.

如果爆破工人需要停留在该区内，他应该在一个安全的位置。

No flow of water occurs through the pipe unless there is a difference in pressure.

除非有压差，否则水是不会流过管子的。

7. 让步状语从句

让步状语从句表示某一动作或状态与另一动作或状态存在意义上的矛盾，但不妨碍事实的进行。翻译时，位置比较灵活，一般放在主句之后，也可根据情况，放在主句之中。

让步状语从句常用的引导词有：although，while，as，though，even of，whether. . . or，no matter what 等。

While the main fan, or combination of main fans, handles all of the air that circulates through the underground network of airways, underground booster fans serve specific districts only.

虽然主扇或风机组合能够处理所有通过航空公司地下网络流通的空气，但是地下增压风机只能为特定地区服务。

Although accidental fires are on the increase, one of the most concerning categories of fires are those related to arson.

虽然意外火灾正在不断增加，其中最受关注的类别之一是涉及纵火的火灾。

Exposure Assessment to Dust and Free Silica for Workers of Sangan Iron Ore Mine in Khaf, Iran

Silica (SiO_2) is one of the most abundant minerals on the earth, extremely toxic, and also, can cause fibrosis. Exposure to silica compounds more than ***permissible*** limits is known as a hazardous potential in pulmonary disease. Also, silica contributes in ***silicosis*** and sclerosis. One of the most important mines in Iran is iron ore mines. However, the high quantities of dust and particles are produced during the digging, stonecutting, and rock drilling processes. Meanwhile, different particles can cause different lung injuries.

Quartz is a term that often refers to free silica dust. Silicosis is a lung disease and decreases the ability of the lungs in oxygen up taking. Overexposure to crystalline silica in either respirable or irrespirable forms can cause adverse health effects. In case of respirable silica health effects, silicosis is one of the most important reported diseases. As a result, more than 250 workers die each year in the United States because of silicosis, and more than hundreds of workers are disabled because of silicosis and bronchitis. Unfortunately, silicosis is not curable, but it can be prevented by reduction and control of exposure to silica compounds. In a study that conducted by Samadi in Emarat Lead mine of Iran, the concentration of respirable ***free silica dust*** was a few times higher than standard level. Moreover, in another study in ferrosilicon mine of Semnan, exposure level to respirable silica dust was evaluated and reported to be higher than National Institute of Occupational Safety and Health (NIOSH) standard level. Also, in another report, 8 h occupational exposure level in the United States during 1998–2003 was reported higher than American Conference of Governmental Industrial Hygienists (ACGIH) standard level. However, during the years 1998–2003, exposure trend was declining, but still it was higher than Occupational Safety and Health Administration (OSHA) standard levels. Exposures in occupational fields are in very wide ranges. For example, occupational exposures occur by working in ore, coal, nonmetallic mineral and stone mines, coarse and fine sand and ***clay*** digging, shingle making, ceramic manufacture, non-iron melting, and so on. Pulmonary disorders due to dust particles are one of the oldest occupational diseases. This kind of pneumoconiosis appear in exposure to free silica from ore digging or crusher station and is a difficult and costly work-related disease.

The International Agency for Research on Cancer (IARC) was classified ***crystalline*** silica as a "group 1 human lung carcinogen" in 1997. Also, current evidence implies that free silica is an effective potential in lung cancer occurrence. Considerate to improved work conditions and dust control in developed countries, incidence rate of silicosis is declining. While in developing countries, exposure to dust is an important health problem. Hence, dust concentration cognizance in workplace air, particularly free silica, has been practiced to minimize the adverse health effects.

Khaf is a county in Razavi Khorasan Province in Iran in neighboring with Afghanistan. It is a small border town about 250 km from Mashhad. The historical city of Nashtifan with centuries-old windmills is located close to Khaf.

Sangan iron ore mine in Khaf is one of the richest mines in the Middle East region. It has been estimated that discovered iron ore deposit be about 1, 200 billion tons that some 500,000 tons exploit each year. Production is foreseen to reach 40. 2 million metric tons in 2013 and 42 million metric tons in 2014.

According to the data gathered from this mine, there is a high silica concentration in extracted stones of this region. Therefore, it is obvious that silica can ***spread out*** in the air during excavation or mining the ground and can cause various dangers for workers, such as silicosis. Thus, this survey was conducted in 2010 in order to evaluate the dust and free silica concentrations in different indoor air of Sangan iron ore mine in Khaf.

Typically, respirable silica has been analyzed via two methods including X-Ray Diffraction (XRD) and infra-red spectrometry. Of these, XRD is the more accurate and less ***susceptible*** to interferences from other minerals that will be found in respirable dust (as defined as less than 10 μm in aerodynamic diameter or PM10).

XRD can also distinguish the three main types of silica, namely as crystobalite, tridymite, and alpha quartz on single 25 mm personal air sampling filter. This is important in foundries and other high temperature environments were quartz can ***denature*** to tridymite and crystobalite at high temperatures, both of which are hazardous chemical substances.

Free silica concentration in total, respirable, and control samples were collected from workplace environment and analyzed based on XRD technique. The total and respirable sampling was accomplished based on 7500 NIOSH method (NIOSH 2003). In order to remove humidity from filters, before weighing, they were maintained and dried for at least 24 h in desiccators, before sampling. Filter weighing was performed by digital balance with precision of 0.001 g. Selected membrane filters had 25 mm diameter and were provided from Sartorius Company (Germany). Sampling pump (MCS-10 model) was provided from SKC Company (the United States). As calibration is required for sampling train, we calibrated the pump using a flow meter before sampling and after each repair or abuse in the field. For calibration of the sampling pump, the suction tube of sampling pump was connected to the inlet of the flow meter. Then, by adjusting the flow rate of the pump on a specific rate, the flow rate was read on the flow meter and was considered as true flow rate of the sampling pump.

Sampling was based on environmental and personal sampling methods. In environmental sampling method, sampling apparatus including filter and filter holder, environmental sampling pump, and flexible pipe were prepared and fixed in sampling stations. Environmental sampling period was 2 h and flow rate was 8 L/min. In personal sampling method, the flow rate of the personal samplers was 1.7 L/min and the pumps stopped automatically at 4 h. Personal sampling apparatus was included cyclone and holder (clung on worker collar), sampling pump (clung on worker waist) and flexible pipe. At the end of the sampling periods, filters were transferred to the laboratory and, after drying in desiccator, weighted with digital balance, thereby total and respirable dust concentration were computed based on the following equation at milligrams per cubic meter (mg/m^3)

$$C = \frac{(W_2 - W_1) \times 10^3}{\Delta t \times Q}$$

where C = dust concentration of air in workplace;

W_1 = filter weigh before sampling, mg;

W_2 = filter weigh after sampling, mg;

Δt = sampling period, min;

Q = sampling pump flow rate, L/min (with correction of sampled air volume to volume in standard condition).

Finally, 96 workers in the mine were examined medically. Environmental samples, control samples, and also 26 standard silica samples were analyzed with XRD technique.

Based on the NIOSH instruction, XRD technique was used for free silica (quartz) determination in dust samples. Considerate to the fact that XRD quantitative analysis technique need standard curve to determine unknown sample concentrations, standard samples should be prepared. Filtration method was used to achieve this purpose. Thus, pure and standard respirable quartz powder samples in 10 and 50 mg prepared. Then, each sample was transferred to a 1 L capacity beaker and filled to the mark with 2-propanol solution. Thereafter, the mixture agitated for 20 min until a suspension was reached. Simultaneously, a silver filter with 25 mm in ***diameter*** and 0.8 lm in pore size diameter was fixed on the holder and a pump with 1.7 L/min was turned on. Based on the simple mathematical computations, appropriate amount of suspension was picked up by a pipette and poured on the silver filter. Eventually, on the several standard filters of 10, 20, 30, 50, 100, 200, 500, and 800 μm, standard silica was existed. After standard sample preparation, they were set into diffract meter in turns and intensity was read. Then, their standard regression curve was drawn. In order to analyze the total and respirable dust in samples, membrane filters were set into specific area of diffract meter and quartz peak intensity was obtained based on cycle per second (CPS) and each filter peak intensity was read. These results were reported as geometric mean and were compared with ACGIH, NIOSH, and Iranian Standards.

In this study we monitored the levels of silica in air of Sangan iron ore mine for distinguishing that the workplace is dangerous for workers or not. Because of the obvious dangers involved with inhaling silica dust, it is important for a monitoring system to be in place to ensure levels of respirable crystalline silica are within 0.1 mg/m^3 for mining and 0.4 mg/m^3 for general industry. These levels are prescribed in the *Hazardous Chemical Substances Regulations Act* of 1995 and the *Occupational Health and Safety Act* 85 of 1993.

In this study, in order to identify whether silica exists in zoisite or not, eight stone and powder samples were collected from different parts of the mine and analyzed by XRD technique (Chen et al., 2010). As a result, high percentage of free silica (mean quartz concentration = 10.8%) was found. Altogether, 48 environmental samples were collected and analyzed by eight work stations including crusher machine, crusher machine loaders, site loading, Tappeh Ghermez drilling No. 1, Tappeh Ghermez drilling No. 2, rest room of Tappeh Ghermez workers, rest room of crusher machine workers, and official and safeguard stations. Volume selections of the environmental samples were based on sampling from outset to the end of the mine, and whole mine stations were sampled.

Eventually, there are many specific and general recommendations to reduce exposure to respirable crystalline silica at the workplace. Workers can limit their exposure by being aware of and practicing the following tips:

- Identify through measurements in the workplace where silica dust may be generated.
- Use containment and/or controls methods, such as wetting dust producing areas. Thereby limiting exposure of workers to hazardous dust.
- Regularly maintain dust control systems to ensure they are working as well as possible.
- Conduct air monitoring to measure exposure of workers.
- Install dust extraction equipment preferably with bag filtration.

The use of ***respirators*** or personal filtration masks should be enforced where dust levels are high.

- Warning signs should be placed at entrance of any area where high levels of dust ***are going to*** be encountered.
- Inform staff about hazards of respirable crystalline silica dust and any other hazardous dust they may encounter in the workplace.
- Issue protective masks and respirators at areas where high dust exposure is going to be encountered.

New Words and Expressions

permissible [pəˈmɪsɪb(ə)l] *adj.* 容许的，许可的

silicosis [ˌsɪlɪˈkəʊsɪs] *n.* 硅肺

free silica dust 自由二氧化硅粉尘

clay [kleɪ] *n.* 黏土，陶土

crystalline [ˈkrɪst (ə) laɪn] n. 晶状体球蛋白
spread out 散开，铺开，展开，摊开，分头行动
susceptible [səˈseptɪb (ə) l] adj. 敏感，易受影响，过敏，好动感情的
denature [diːˈneɪtʃə] v. 使变性，使（酒精）不能饮用，使变质
diameter [daɪˈæmɪtə] n. 直径，对径，放大率，放大倍数
respirator [ˈrespɪreɪtə] n. 口罩，防毒面具
be going to 将要，打算

Unit Eleven

Construction Safety

Construction safety is a global issue in that it is a concern wherever construction activities ***take place***. The reality is that the construction industry continually has injury and ***fatality*** statistics that make it one of the most dangerous industries in which to work. Even though tremendous improvements have been made in safety performance in some countries, the construction industry continues to lag behind most other industries. This has been the experience within most countries. As the world has become smaller through technology and through cooperative arrangements that cross many borders, the issue of construction worker safety has become a well-recognized problem and represents a concern that is shared worldwide.

The construction safety problems that exist are rarely unique to a single country. In the global market, construction problems are very similar from country to country. This is quite evident when attending international construction safety conferences where the themes of primary interest have general appeal to all participants. Since construction safety problems appear to be ubiquitous, this also means that the problems of construction safety can be addressed and solved on a global scale, resulting in improvements that can be observed on a global scale. Thus, solutions to safety problems in one country can readily be adopted in other countries to generate further improvements.

Professional engineers ***are dedicated to*** protecting the public safety, and many of our members are in positions of "trust" in our nation's building departments where they struggle to provide codes that minimize the public's exposure to accidental injury. The problem is twofold in that the written applicable regulations, codes, standards, and safety rules, such as those provided by government OSHA, industry crane manufacturers, and professional societies provide the structure while enforcement and training must support project installation procedures. Construction engineering practitioners are facing a dilemma in safeguarding workers and the community affected by a building project. These engineers, who represent owners through their managers and contractors, have installed a series of safe standards that generally lead to a low injury rate on a project; owners must participate actively in project safety including the assignment of ***contractor*** responsibility to prepare a site safety plan that includes provisions for ***crane*** and hoisting safety. The owner must require compliance with the applicable regulations, codes, and standards.

"The devil is in the details," and the engineering community is being held liable for many injuries that are caused by human error. Recent examples include a crane accident possibly caused by faulty nylon straps that failed due to either overuse, misuse, or abuse to temporarily support the six-ton bracing collar during the jump of a 200-foot-tall tower crane. OSHA has announced that they will release a proposed revision of its crane and derrick standard. Having a crane manufacturer representative, a master ***rigger***, and a safety specialist on site during a jump may help. Adding layers of supervisors may also be ***counterproductive***; the responsible parties must educate their employees and supply them with safe tools and supports. The architect/engineer, as action agent for the owners' contract documents, has a responsibility to include contract requirements that specify regulations, codes, standards, and sequences such as ASCE's policy statements on construction site safety, crane safety, and their crane safety manual, while the remainder of the team is responsible for implementing the proper steps in accordance with the required specifications. Another example is the case where a crane's swing table separated from its ring support. Earlier, a cracked part had been repaired by welding and reinstalled; the catastrophic failure of the crane occurred while operating. The public must be protected during crane and hoisting operations.

The suggestion that employers observe workers and provide feedback to them on both safe practices and at-risk actions has been used on some sites and is known as behavioral-based safety. Anonymous safety records for workers that are transferable to other sites and contractors may be a method of improving safety awareness provided that it does not add to the paper backlog. It seems like each suggested solution has many negatives that are not desirable. The responsibility for worksite safety must be assigned to all the team members, which includes government and private owners, insurers, designers, and contractors. The worker that steps on an unsafe scaffold—one that he or she built—will often seek damages from the entire project participants. Perhaps, if a special safety policy group were set up on the site to monitor safety actions and damages then projects could avoid some of the pitfalls that lead to serious accidents. Cameras are common on a construction site to record daily progress; they should be upgraded to record much of the site work activity in detail. Each accident, no matter how small or large, should be investigated to learn how to avoid it in the future. In such cases a film of the activity in the vicinity of the accident would be a great help. The owner must assign safety as a responsibility to his primary contractor authority and this agent must coordinate ***site safety*** activities, including the management of crane and hoisting operations.

A number of serious failures have occurred in steel structures, including the Minneapolis I-35W bridge disaster. Connections are the most common cause of claims and failures and they are the most difficult part of steel design. Many codes assign responsibility for their suitability to the structural engineer, though it is standard practice for the structural engineer of record to designate the connection design to a professional engineer working for the steel fabricator. In most cases the structural engineers conduct peer reviews of their work in accordance with recommendations of various structural associations and as required by many insurance companies representing the structural design engineer. Building codes are being revised in an effort to increase a building's resistance to progressive collapse through stronger horizontal ties. Mandated peer reviews may correct some of these potential hazards.

Progressive failures that result from destructive fires leave unanswered many questions regarding the speed and extent of the damage in a steel-framed building. The fireworks display ignited the flammable insulation layer after penetrating the exterior metal facade at several points. It then spread to the interior of the building where debris and decorations provided fuel for accelerating the flames. In addition fireworks landing on the roof spread fire to the lower floors where high winds swept it further to debris piles that ignited very quickly. Chimney effects in the hollow core between the 5th and 25th floors probably accelerated the fires as they spread out from the central core. The heat was so severe that it partially collapsed the core during its passage. It is unknown whether the building will be torn down or if renovations can save the structure's steel skeleton that is still standing.

As engineers representing either government, owners, designers, or contractors, we must work together to prohibit the stockpiling of debris on construction sites and consider the effects of an unexpected fire on a project. We must minimize the use of materials that are flammable and carefully detail the fireproof closure of shafts or cores during a fire emergency while the project is under construction or during its intended life. Perhaps we should consider revising our building codes and require that the sprinkler system be pressurized and operable as early as possible in a project's construction.

Failures that lead to serious injuries have a common thread that passes from crane accidents to ***scaffold*** failures to steel connections and to all kinds of construction accidents. The chain is no stronger than its weakest link, and to strengthen the weak links we must as a profession mandate the use of cameras and the resulting photographic record to instruct workers on what appears to be currently unsafe practices, to encourage OSHA to issue updated regulations, and push employers to train their associates—such as the field representatives of the contractor and of the design engineer—using the most current safety requirements.

Supervisors must audit and check the jobsite procedures during crane jumps. Riggers must follow the manufacturers' instructions and not take shortcuts: In one crane failure the jump crew used half of the eight recommended attachment points and used synthetic slings at an angle that led to an unorthodox choker, decreasing holding capacity and exposing them to sharp edges. The consequences of taking shortcuts on a construction site may lead to disaster. Construction teams and crews do not always perform as expected; qualified rigging professionals must inspect each step. Registered professional engineers should approve the manufacturers' instructions and the overall plans for the construction site. The owner's safety agent must develop a site-specific crane and rigging safety plan, which includes production and critical lifts, and this safety authority must require the training of management staff and jobsite personnel in crane and rigging safety procedures.

Construction work crews must be well-trained in safe practices and engineers must continue to require inspection procedures that mimic jobsite conditions whenever the public is at risk. Engineer approval should be based on test lifts that are similar to jobsite conditions and utilize the site equipment. We, as engineers representing government, owners, designers, or contractors, must do more to overcome careless actions on construction projects. The engineers' vow to protect the public safety requires recognition that having the best system of construction codes and standards does not prevent careless or inconsiderate construction participants from finding a way to introduce unsafe practices, such as contacting electrical power line ***hazards*** on a project or an adjacent site.

Experimental data indicates that employees learn more when they are required to participate in interactive problem solving, like installing a model crane jump or a scaffold. Crews that were required to build devices after having studied the standards, including the manufacturers' installation instructions, demonstrated a better grasp of the issues than the ones who learned only through lectures, reading, and tests. Contractors are encouraged to develop a site-specific safety plan and job hazard analysis for common field activities undertaken by their construction crews. Each craft has risky exposures that may be avoided through practice on safety models. OSHA has developed safety-training courses for both craft workers and managers that should be helpful on most projects. Manufacturers must standardize load chart formats and equipment control configurations with all manuals written for the end user. Documentation should emphasize equipment limitations for wind and water operation.

The major risks associated with crane and hoisting operations demand an increased effort to improve crane and rigging safety by the construction team.

New Words and Expressions

take place	举行，发生
fatality [fəˈtælɪti]	*n.* 死亡，宿命，致命性，不幸，灾祸
dedicate to	奉献，从事于，献身于
contractor [kənˈtræktə]	*n.* 订约人，承包人，收缩物
crane [kreɪn]	*n.* 鹤，起重机，伸长，探头，迟疑，踌躇
rigger [ˈrɪgə]	*n.* 索具装配人，机身装配员，暗中操纵者，脚手架
counterproductive [ˌkaʊntəprəˈdʌktɪv]	*adj.* 反生产的，使达不到预期目标的
site safety	工地安全
scaffold [ˈskæfəld]	*v.* 搭脚手架于（某处），使站在脚手架上，把……处死刑
supervisor [ˈsjuːpəvaɪzə]	*n.* 监督人，[管理]管理人，检查员
hazard [ˈhæzəd]	*n.* 危险，冒险，冒险的事

Exercises

1. What is the status quo of Construction Safety around world?
2. What is the problem for Construction Safety? Explain each of them.
3. What is the reason for crane accident? How do participants do to ensure construction safety?
4. How to prevent fire as engineers in worksite?
5. What should work crews and engineers do to reduce the major risks associated with crane and hoisting?

科技英语翻译技巧——名词性从句

科技英语中的名词性从句有主语从句、宾语从句、表语从句和同位语从句。它们在句子中的功能相当于名词或者名词性短语。

连词：that，whether，as if。

连接代词：what，who，which，whose，whatever，whoever，whomever，whichever。

连接副词：where，when，why，how，wherever，whenever。

1. 主语从句

做主语的从句叫作主语从句。主语从句一般有两种形式：一种是直接将主语从句放在句首，关系代词有 that，what，whatever，which，whoever，whether 等；另一种是 it 做形式主语放在句首，实际的主语放在后面。

1）以 what，who，whether，that，when，where，why，how 等代词引导的主语从句在翻译时可按照原文顺序进行翻译。

What potential energy and kinetic energy really mean will be the content discussed in the next article.

势能和动能的含义究竟是什么将在下文中予以讨论。

Whether there is life on the moon is no longer a question.

月球上是否有生命已经不再是个问题了。

Energy is what brings changes to materials.

给物质带来变化的是能量。

2）形式主语 it + 谓语 + that/whether 引导的从句，翻译时先翻译从句，再翻译主句。

It is a common practice that another worker assists the injured worker in obtaining medical treatment.

这是另一种常见的做法：另一名工人协助负伤的工人获得医疗救治。

It is estimated that fires that are started deliberately account for 40% of workplace fire.

据估计，蓄意造成的火灾占工作场所火灾的 40%。

It is essential that the person carrying out any inspection is sufficiently independent and impartial to allow them to make objective decisions, and has appropriate and genuine authority to discard defective lanyards.

至关重要的是，进行检查的人员要有足够的独立性和公正性，使他们能做出客观的判断，并有适当的和真正的权利，可更换有缺陷的吊带。

2. 宾语从句

宾语从句在修饰词之后，可以做动词（谓语动词和非谓语从句）的宾语，也可以做某些形容词或者介词的宾语，翻译时按照顺序法翻译。

Scientists have long predicated that computers would one day help speed up the arduous task of translating texts.

科学家们早已预言，总有一天计算机将帮助我们加快艰巨的文字翻译工作。

After years of studying pictures from Mariner 9 and Viking orbiters, scientists gradually came to conclude that many features they saw suggested that Mars may have been warm and wet in an earlier era.

科学家们在多年研读“水手 9 号”和“海盗号”探测器发回的照片后，逐渐得出结论，他们所看到的很多特点表明，火星在较早时期可能是温暖且湿润的。

However, we find that the culture and style of management is even more significant, for example a natural, unconscious bias for production over safety, or a tendency to focusing on the short-term.

但是，研究发现管理文化和管理风格更为重要。例如，管理层的一个自然的、无意识的偏见就是，认为生产胜过安全，或者倾向于注重短期的利益。

3. 表语从句

表语从句是在系动词后面的一个由连接词引导的句子，翻译时一般按照原句的顺序。

One of the remarkable things about it is that the electromagnetic waves can move through great distances.

其值得注意的一点之一就是电磁波能够远距离传播。

Perhaps the most common classification of material is whether the material is metallic or non-metallic.

也许，材料最常见的分类是根据它是金属还是非金属来进行的。

The result of invention of steam engine was that human power was replaced by mechanical power.

蒸汽机发明的结果是机械力代替了人力。

4. 同位语从句

同位语从句不同于定语从句，从句既说明抽象名词，又等同于抽象名词，如果是 that 引导的同位语从句，that 不做成分，一般不能省略。定语从句修饰限制先行词，关系词替代先行词在从句中的成分，关系词充当宾语时，可以省略。

The only times that these costs are quantified on a regular basis are in various courts of law when lawsuits are resolved.

只有在各种法律法庭中，为了解决诉讼时，这些成本才会被定期地量化。

The operation of scanning is greatly simplified by the fact that the earth rotates on its axis, so that a fixed aerial upon the surface of the earth will scan a complete circle of the sky in the course of a day.

由于地球围绕着自己的轴线转动，固定在地球表面的天线能在一昼夜之间扫描整个天空，这一情况把扫描的操作过程大大简化了。

Einstein came to the conclusion that the maximum speed possible in the universe is that of light.

爱因斯坦得出这样的结论，即宇宙中的最大速度是光速。

Construction Accident Prevention Techniques

Accident Prevention Techniques (***APTs***) are tools which can be used as an integral part of any safety and health program. Each technique serves a unique function. It will be your job to decide which techniques to incorporate into your safety and health effort.

The first types of APT you may want to use are safety and health management, safety communication, toolbox talks, and training. These techniques set a foundation for accident prevention. Once established, ***identify*** the hazards and plan procedures for avoiding those hazards. Recognize the need to call on a specialist, engineer, expert, or to have some form of a special program to address your company's specific needs (e. g., maintenance program, eye protection program, etc.). Finally, ***evaluate*** your efforts using safety and health audits, Job Safety Observations (JSO), and accident investigations.

This passage explains each technique and denotes its use, including positive and negative aspects. For specific problems, remember that using these techniques as stand-alone entities may have only a band-aid effect for specific problems. The ***implementation*** of a complete approach is a huge step in assuring a functional safety and health initiative. This completeness requires a written program that addresses workers' needs and concerns, hazard recognition and analysis, and compliance with OSHA standards and requirements. Only after you have an organized safety and health program can you expect to effectively apply the APTs listed in this passage; they are only a part of a complete safety and health program. Using this type of approach should afford a workplace free of hazards, worker injuries, and illnesses.

When a construction contract/project is in the implementation stage, the prime contractor/owner needs to address safety and health. This assures that all general contractors and subcontractors comply with the project's job safety and health requirements. Use of the tools presented in this passage will help meet those requirements.

1. Safety and Health Management

Where does safety and health on a construction site/project begin? The answer is, of course, from day one. When does it stop? The answer is "never". Safety and health management should be considered from the design ***inception***, through the bid process, the contract award, and the initial start. Then the management of safety and health continues through the project period and becomes the foundation for the next and all other ensuing projects that your company undertakes.

The phrase "safety and health management" needs a definition which is both practical and relevant to the construction industry. Thus, in the broadest sense, safety and health management means taking control of the existing hazards. Develop a systematic or programmed approach which allows you to effectively, efficiently, and productively direct, mesh, control, and intervene in a complex set of variables and contributing factors (e. g. , risks, hazards, job tasks, other contractors, employees, machines/equipment, or the environment). These factors may impact your attempts to complete a project in a safe, healthy, and responsible manner.

2. Communications

Communication is the key to occupational safety and health. The message of accident prevention must be constantly reinforced. Use constant reminders in the form of message boards, fliers, newsletters, paycheck inserts, posters, the spoken word, and face-to-face encounters. Make the message consistent with the policies and practices of the company so it is believable and credible. It is important to communicate safety and health goals and provide feedback progress toward accomplishing those goals. Most workers want their information in short, easily digested units.

Face-to-face interactions personalize the communications, provides information immediately (no delays), and allows for two-way communications which improves the accuracy of the message. Finally, these interactions provide for performance and real-time feedback and reinforcement regarding safety and health.

3. Toolbox Talks

Toolbox safety talks are especially important to the supervisors on your job sites and projects because they afford the supervisor the opportunity to convey, in a timely manner, important information to the workers. Toolbox talks may not be as effective as the one on one, but still surpass a ***memorandum*** or written message. In the five to ten minutes prior to the workday, a shift, at a break, or as needed, this technique helps communicate time-sensitive information to a department, crew, or work team.

In these short succinct meetings, convey changes in work practices, short training modules, facts related to an accident or injury, specific job instructions, policies and procedures, rules and regulation changes, or other forms of information which the supervisor feels are important to every worker under his supervision.

Although toolbox talks are short, these types of talks should not become just a routine part of the workday. Thus, in order to be effective, they must cover current concerns or information, ***be relevant to*** the job, and have value to the workers. Carefully plan toolbox talks to effectively transmit a specific message and a real APT. Select topics applicable to the existing work environment, plan the presentation and focus on one issue at a time. Use materials to reinforce the presentation and clarify the expected outcomes.

Some guidelines are as follows:

- Plan a toolbox training schedule in advance and post a notice.
- Prepare supporting materials in advance.
- Follow a procedure in the presentation: explain goals, try to answer questions, restate goals, and ask for action.
- Make attendance mandatory.
- Make each employee sign a log for each session.
- Ask for feedback from employees on the topic or other proposed topics.
- Involve employees by reacting to suggestions or letting them make presentations when appropriate.
- Reinforce the message throughout the work week.

No matter how effectively you communicate with your workforce, you still need to assure that your workforce has the competence to perform the basic skills for the tasks they have been assigned.

4. Training

You may offer all types of programs and use many of the APTs, but, without workers who are trained in their trade and have safe work practices, your efforts to reduce and prevent accidents and injuries will result in marginal success. If a worker has not been trained to do his or her job in a productive and safe manner, a very real problem exists.

Do not assume that a worker knows how to do his or her job and will do it safely unless he or she has been trained to do so. Even with training, some may resist safety procedures and then you have a deportment or behavioral problem and not a training issue.

It is always a good practice to train newly hired workers or experienced workers who have been transferred to a new job. It is also important that any time a new procedure, new equipment, or extensive changes in job activities occur, workers receive training. Well-trained workers should be more productive, more efficient, and safer.

Training for the sake of documentation is a waste of time and money. Training should be purposeful and goal or objective driven. An organized approach to on-the-job safety and health will ***yield*** the proper ammunition to determine your real training needs. These needs should be based on accidents/incidents, identified hazards, hazard/accident prevention initiatives, and input from your workforce. You may then tailor training to meet the company needs and that of workers.

Look for results from your training. Evaluate those results by looking at the number of reduced accidents/incidents, improved production, and good safety practices performed by your workforce. Evaluate the results by using JSO and safety and health audits, as well as statistical information on the numbers of accidents and incidents.

Many OSHA regulations have specific requirements on training for fall protection, hazard communication, hazardous waste, asbestos and lead abatement, scaffolding, etc. It seems relatively safe to say that OSHA expects construction workers to have training on general safety and health provisions, hazard recognition, as well as task-specific training. Training workers regarding safety and health is one of the most effective APTs.

5. Hazard Identification

In order to prevent accidents, identify existing and potential hazards which can prevail upon the worksite. When looking at specific jobs, break the job down into a step-by-step sequence and identify potential hazards associated with each step. Consider the following questions:

1) Is there a danger of striking against, being struck by, or otherwise making injurious contact with an object? (E. g. , being struck by a suspended drill casing or piping as it is moved into place.)

2) Can the employee be caught in, on, or between objects? (E. g. , an unguarded v-belt, gears, or reciprocating machinery.)

3) Can the employee slip, trip, or fall on same level, or to another level? (E. g. , slipping in an oil-changing area of a garage, tripping on material left on stairways, or falling from a scaffold.)

4) Can the employee strain himself or herself by pushing, pulling, or lifting? (E. g. , pushing a load into place or pulling it away from a wall in a confined area.) Back injuries are common in every type of industrial operation; therefore, do not overlook the lifting of heavy or awkward objects.

5) Does the environment have hazardous toxic gas, vapors, mist, fumes, dust, heat, or ionizing or nonionizing radiation? (E. g. , arc welding on galvanized sheet metal produces toxic fumes and nonionizing radiation.)

Review each step as many times as necessary to identify all hazards or potential hazards. With the identification of the hazards, take steps to prevent the accidents or incidents from occurring. If you know the hazard, it is easier to develop interventions which mitigate the risk potential. These interventions may be in the form of safe operating procedures.

At times a hazard hunt is appropriate. This constitutes providing a form (see Table 11-1) to your workers and asking them to list anything which they perceive to be a hazardous condition on the job site. You will be able to review these hazard hunt forms to determine if hazards truly exist or are just perceived. If you use the hazard hunt as a hazard identification process, it is important to provide feedback to the workers on whether a hazard does or does not exist. It is never wise to disregard this important communication. Hazard identification has always been an effective APT.

Table 11-1 Hazard Hunt Form

Worker's Name (Optional): Date: Jobsite: Job Title:

（续）

1. Describe the hazard that exits.
2. Has the hazard been discussed with your supervisor? When? (Date)
______Yes ______No
3. What preventive action needs to be taken?
4. Do you know what state or federal safety regulations is being violated?
______Yes ______No
(Please list it if your answer is yes.)
5. Have there already been injuries or illnesses caused by the hazard?
______Yes ______No
(List the names, dates, events, or symptoms.)
6. How long has the hazard been present?
Manager's or Supervisor's Response to Hazard Concern Identified.
NOTE: Use a Separate Form for Each Hazard Identified.

6. Accident Investigation

It is important to have some mechanism in place to analyze accidents/incidents to determine the basis of cause-and-effect relationships. You may determine these types of relationships only when you actively investigate all accidents and incidents which result in injuries, illnesses, or property/equipment/machinery damage. You must have a system in place for these investigations.

Once the types of accidents/incidents which are transpiring have been determined, you can undertake prevention and intervention activities to assure that there will be no recurrences. Even if you are not experiencing large numbers of accidents/incidents, you still need to implement activities which actively search for, identify, and correct the risk from hazards on job sites. Reasons to investigate accidents/incidents include the following:

1) To know and understand what happened.

2) To gather information and data for present and future use.

3) To deter cause and effect.

4) To provide answers for the effectiveness of intervention and prevention approaches.

5) To document the circumstances for legal and workers' compensation issues.

6) To become a vital component of your safety and health program.

If there are only a few accidents/incidents, you might want to move down one step to examine near misses and first aid-related cases. It is only ***a matter of*** luck or timing that separates the near miss or first-aid event from being a serious, recordable, or reportable event. The truth is you probably have been lucky by seconds or inches (A second later and it would have hit someone or an inch more and it would have cut off a finger). Truly, it pays dividends to take time to investigate these accidents and incidents occurring in the workplace.

New Words and Expressions

Accident Prevention Techniques (APTs)　　事故预防技术
identify [aɪ'dentɪfaɪ]　　v. 鉴定，确认，发现，找到

evaluate [ɪ'væljʊeɪt] *v.* 评价，评估，估计
implementation [ɪmplɪmen'teɪʃ(ə)n] *n.* 执行，落实
inception [ɪn'sepʃ(ə)n] *n.* （机构、组织等的）开端
memorandum [memə'rændəm] *n.* 协议备忘录，建议书，报告
relevant to 和……相关
yield [jiːld] *v.* 屈服，让步，放弃，提供
a matter of 大约，……的问题

Unit Twelve

12

Risk Assessment and Risk Management

1. Developing Risk Acceptance Criteria

Before any analysis is done, management should establish ranges of risks that are acceptable and unacceptable, which risks require further investigation, and those that require management review. Management should also make sure the definitions of risk and its components (the hazard and the hazardous situation) are clearly defined in the engineering standards and understood by the entire chain of command. This understanding is critical for clear communications.

Risk is defined as the combined impact of the severity of harm and the probability of harm. Therefore, we need acceptance criteria for each combination of severity and the probability of harm. This is done by developing a hazard assessment matrix, also called a risk assessment ***matrix***. An example is shown in Figure 12-1. Each cell represents a combination of severity and probability of harm and is usually color coded to show if the risk is acceptable, unacceptable, or needs further investigation or management review. Red is usually used for unacceptable risks, green for acceptable risks, and yellow where management review is required.

2. Risk Analysis Using PHA

Before we approve the specifications, we must ***brainstorm*** for all possible ways a patient can be harmed by the medical device and mitigate all major risks by designing them out. Fortunately, a technique—widely used in aerospace, aviation, and commercial transportation industries for decades—exists. This technique is called ***Preliminary Hazard Analysis*** (***PHA***). It covers all types of unsafe events, including ***sentinel***, adverse, and never events. This is the first risk analysis technique in the ISO 14971 standard.

The Joint Commission on Accreditation of Healthcare Organizations (JCAHO) defines a sentinel event as an unexpected occurrence involving death or serious physical or psychological injury, or the risk thereof,

Frequency of Occurrance	Hazard Categories			
	1 Catastrophic	2 Critical	3 Marginal	4 Negligible
(A) Frequent	1A	2A	3A	4A
(B) Probable	1B	2B	3B	4B
(C) Occasional	1C	2C	3C	4C
(D) Remote	1D	2D	3D	4D
(E) Improbable	1E	2E	3E	4E

Unacceptable risk: 1A, 1B, 1C, 2A, 2B, 3A
Unacceptable (investigate risk reduction): 1D, 2C, 2D, 3B, 3C
Acceptable with management review: 1E, 2E, 3D, 3E, 4A, 4B
Acceptable risk without approval: 4C, 4D, 4E

Figure 12-1 An Example of Risk Assessment Matrix

such as harm caused by a malfunctioning surgical robot. Serious injury specifically includes loss of limb or function. The JCAHO defines an adverse event as an untoward, undesirable, and usually unanticipated event, such as the death of a patient, an employee, or a visitor in a healthcare organization. Harmful incidents such as those resulting from improper use of an infusion pump, or from its improper programming by the caregiver, or from an improper design of the software in the medical device, are also considered adverse events even if there is no permanent harm to the patient. The National Quality Forum (NQF) defines never events as errors in medical care that are clearly identifiable, preventable, and serious in their consequences for patients, and those that indicate a real problem in the safety and credibility of healthcare. Examples of never events include surgery on the wrong body part, a foreign body left in a patient after surgery, a mismatched blood ***transfusion***, picking a wrong dose from the drop-down menu on the computer, a severe pressure ulcer acquired in the hospital, and preventable postoperative deaths.

The PHA process consists of the following:

- identifying the potential hazards in the device, its use, and foreseeable misuse;
- assigning the criticality based on the probability of the harm and severity for each hazard;
- performing risk assessment;
- identifying mitigation strategies.

To start, we need to do some research on adverse, sentinel, and never events that are already known in the industry. Such data can be found on the websites of the Joint Commission, the Agency for Healthcare Research and Quality (AHRQ), the Leapfrog Group, the US Food and Drug Administration (FDA), and Consumers Union. The Joint Commission, for example, includes more than 800 sentinel events on its web-site. Consolidate the data relevant to your product and compile a list of hazards that lead to such events. Remember that your device may work perfectly, but if it causes hazards in connected devices, then it can be a hazard related to your device. Add hazards from your own organization's experience, lessons learned, incident reports, and complaints from the families of patients. Such a list is called a ***preliminary*** hazard list. It is just a master list of hazards. In addition, cross-functional brain-storming should be done to add hazards that could occur but are not on the list. The next step is to assess the risk for each hazard in terms of the severity and the probability of the risk. The final step is to specify mitigation actions (see Table 12-1).

Table 12-1 Preliminary Hazard Analysis

System: _ Accu-check _ Meter ___ Date _____ Team Members:					
Item Number	**System Component**	**Hazard Description**	**Hazard Effect**	**Severity Probability**	**Mitigation Strategy**
1	Test strip	Customer using expired strip	Erroneous sugar readings that may result in automotive accidents	Ⅰ/D	Ask supplier to give warning to users in bold print on the container
2	Glucose meter	False positive-sugar too high	User may take insufficient insulin, go into coma	Ⅱ/C	Design the meter for higher accuracy
		False negative-sugar too low	User may take too much insulin	Ⅱ/C	Design the meter for high accuracy or warn users to make frequent checks
3	Control solution to verify the meter accuracy	User ignores need to verify accuracy	User may take too much medication or too little medication	Ⅰ/B	Current instruction sheet is inside the box. Many users find it inconvenient to read. Put the instructions on the box

Note: Severity legend: Ⅰ-Catastrophic, Ⅱ-Major, Ⅲ-Marginal, Ⅳ-Negligible.
Probability legend: A-Very High, B-Moderate, C-Low, D-Very Low, E-Remote.

3. Assessing the Risk

Risk assessment does not have to be a statistical process; it can be done qualitatively. Risk is the function of the severity of an event and the probability of harm. In the PHA in Table 12-1, the qualitative values are used for both. The fifth column in this table shows roman numerals for severity rating and alphabetical descriptions for probability ratings (A for the highest probability, and E for the lowest probability). There are four qualitative ratings for severity (some companies have three or five depending on their choice):

Ⅰ—Catastrophic Harm: Potential for death.

Ⅱ—Major Harm: Potential for disabling harm or serious long-term illness.

Ⅲ—Marginal Harm: Requires medical intervention for a short time.

Ⅳ—Negligible Harm: Does not require medical intervention.

The probability scale is A through E. The meaning for each should be determined by each company. Examples are as follows:

A. Frequent: Can happen once a month or more frequently.
B. Probable: Can happen once a year.
C. Occasional: Can happen once in three years.
D. Remote: Can happen once in five years.
E. Improbable: Should never happen, but is possible.

If we look at the first row in Table 12-1, the consequence of an erroneous reading of sugar level can cause a fatality if the user drives a car unsafely with a high sugar level in the blood. Therefore, the severity rating is I. The chance of such a harm happening may be about once in five years, resulting in the probability rating of D. The combination of severity and probability is called criticality, which is I/D for this hazard. In other words, there is a remote chance that a fatality can happen. This is the statement of risk assessment.

4. Mitigating Risks Using World-Class Practices

The ISO 31000 standard is the generic risk treatment standard at the business level, and suggests selection of one or more of the following options for managing risks, and implementing those options. Risk treatment options are not necessarily mutually exclusive. The options can include combinations of the following:

Risk Transfer: Sharing the risk with another party or parties (including suppliers and insurers).

Risk Avoidance: Avoiding the risk by deciding not to start or continue with the activity that gives rise to the risk or removing the risk source.

Risk Mitigation: Changing the likelihood; changing the consequences (impact).

Accept the Risk: Retaining the risk by informed decision; taking the risk to pursue an opportunity.

Selecting the most appropriate risk treatment option involves balancing the costs and efforts of implementation against the benefits derived, with regard to legal, regulatory, and other requirements such as social responsibility and protection of the natural environment.

A number of treatment options can be considered and applied either individually or in combination. Risk treatment itself can introduce new risks. A ***significant*** risk can be the failure or ineffectiveness of the risk treatment measures. Therefore, monitoring needs to be an integral part of the risk treatment plan to give assurance that the measures remain effective.

5. Risk Evaluation

The purpose of ***risk evaluation*** is to assist in making decisions when the outcomes of risk analysis are questionable and no more mitigation seems feasible. In some cases, the risk is not acceptable but the benefits to the public outweigh the risks. Decisions should be made considering legal, regulatory, and other requirements. Of course, the FDA must approve the device.

In some circumstances, the risk evaluation can lead to a decision to undertake further analysis. There is always a better way.

6. Managing Residual Risks

After we do everything right, customers are likely to discover new risks. These were either overlooked or the operating environment was not fully understood. Typical sources of this information are as follows:

- nonconforming reports on suppliers;
- nonconforming reports in production;
- incidence reports in hospitals;
- clinical data;
- post-market surveys;
- complaints;
- medical device records;
- technical conferences.

New Words and Expressions

matrix [ˈmeɪtrɪks] n.. 矩阵，基体，社会环境，线路网

brainstorm [ˈbreɪnstɔːm] v. 集体讨论，集思广益以寻找

Preliminary Hazard Analysis 危害分析，预先危险性分析

sentinel [ˈsentɪn(ə)l] n. 哨兵

transfusion [trænsˈfjuːʒ(ə)n] n. 追加投资，(资金的) 注入

preliminary [prɪˈlɪmɪn(ə)ri] adj. 预备性的，初步的，开始的

risk mitigation 风险消减，风险缓解

significant [sɪgˈnɪfɪk(ə)nt] adj. 有重大意义的，显著的，有某种意义的，别有含义的

risk evaluation [经] 风险评价，火险评估

Exercises

1. What is risk acceptance criteria? What kinds of risk are divided in the Hazard Categories?
2. What is PHA? And what does the PHA process consist of?
3. How many qualitative ratings for severity are there and what are they?
4. Explain the definition of risk evaluation? What is the function of risk evaluation?
5. What are typical sources of risk assessment and risk management information?

科技英语翻译技巧——被动句

科技英语中叙述的对象往往是现象、规律、过程等，需要强调的是所述客观事实本身，而不是行为主体是什么。被动句的表达更为清楚客观，强调了作者所表达的客观现象，这种句式更能引起读者对作者所述规律、现象的理解，因此被动句在科技英语中普遍应用。

1. 译成主动语态

在汉语中，普遍习惯使用主动句，这样更符合汉语的行文规范，更方便读者理解。在翻译时，通常会将科技英语中大量的被动句翻译成主动句。

被动句有如下几种形式：

1）突出动作者或强调某种行为或动作上的结果，将承受者或结果采用倒叙的方式，放在主语

的位置上，使句子的表达更为清晰。这种句子并不是被动的意义。

2）突出某种状态或结果，但行为者不明确。这种句子也不是被动的意义。

3）由 by 引导行为主体，句子有被动的意义。

翻译被动句时，将被动句转化为主动句，有如下几种方法：

（1）原文主语不变

不改变句子的结构，将“被”字省略，直接翻译成主动语态。

Solution to the air pollution from dust particles or gases *was ultimately found.*

解决防止灰尘或煤气污染大气问题的方法终于找到了。

Early fires on the earth *were certainly caused* by nature, not by man.

地球上早期的火肯定是由大自然而不是人类引燃的。

Other questions will *be discussed* briefly.

其他问题将简单地加以讨论。

Nuclear power's danger to health, safety, and even life itself can *be summed up* in one word: radiation.

核能对健康、安全，甚至对生命本身构成的危险可以用一个词——辐射来概括。

（2）主语译成宾语

把句子的原主语翻译成宾语，而把行为主体或相当于行为主体的介词宾语翻译成主语。

Friction can be reduced and the life of the machine prolonged by lubrication.

润滑能减少摩擦，延长机器寿命。

All the wire would get red-out with such a current, and fire would be caused.

这样的电流会使所有导线炽热，从而引起火灾。

（3）增加主语

1）在翻译某些被动语态时，为了使句子更加通顺，需要增加例如“我们”“人们”“大家”等这样的主语，然后将原句的主语译成宾语。

Design changes are then made to minimize these risks.

为了减小这些危险性，（人们）随后做出了设计上的更改。

2）由 it 做形式主语的被动句，翻译时按主动句翻译。在翻译时，将 it 作为一个独立的部分，而将主语从句放在宾语的位置上。常见句型有：

It is well known...

It is believe...

It is generally considered...

It will be said...

It would be told...

It is possible that work procedures must be changed or that maintenance is needed.

工作程序需要改变或者进行维护，（人们）认为这是有可能的。

3）在有复合宾语的动词用于被动语态时，如“believe”“consider”“know”“see”“think”等，翻译时加不确定主语。

It is *believed* that a new technology will restore the particulate matter level of the atmosphere to that of a century or more ago.

我们相信一项新的技术将会使大气中颗粒物恢复到约一个世纪以前的水平。

2. 译成汉语的其他句型

（1）译成汉语的无主句

翻译时，不把原句的主语翻译成汉语中的主语，又不另加主语，可将句子翻译成汉语中的无主语形式。

The fuel *must be tested to* determine its suitability before application.

使用前须对燃料进行试验，以确定其适用性。

（2）译成判断句

描写事物的过程、性质和状态的句子，可以翻译成“是……的”判断句。

Accidents are caused by indirect and direct causes.

事故是由间接原因和直接原因造成的。

（3）保持被动结构

在英语中，被动句的标志是“be + 过去分词”，而在汉语中，被动句的标志是“被……”等具有被动含义的词。在翻译被动句时，有时还需保持原有的被动语态。使用“被”“由”“受”“为……所”“使”“把”等词。

The hypothesis has *been proved up* to the hit by the results of experiments.

这一假说已经被实验结果完全证明。

Methods and Models in Process Safety and Risk Management: Present and Future

1. Origin of Concept of Process Safety and Driving Forces for Its Development

The origin of the term “process safety” and its international evolution is associated with the major process accidents that occurred during the time period between 1960 and 1990. It is evident that the concept of process safety has been successfully used in process industries even before this time period. E. I. DuPont, founded in 1802 to manufacture gunpowder, has successfully utilized the concept of process safety to prevent serious injuries and incidents creating a good foundation for current process safety and risk management (Klein, 2009). More importantly, the contribution of Professor Trevor Kletz to the field of process safety and his involvement with the origin and evolution of process safety is worthy to discuss throughout this article. The evolution of process safety is closely connected with Professor Trevor Kletz's professional and academic career, especially his professional career at Imperial Chemical Industries (ICI), England (Kletz, 2012). Scientific research on process safety and risk also started ***simultaneously*** and it is considered that the 1970s were the golden decade of research in this field.

2. Current Research Trends

The future trend of process safety and risk management development inclined to the following areas:

- hazard identification and analysis;
- risk assessment;
- safety management;
- inherent safety.

(1) Hazard Identification and Analysis

1) Atypical hazard identification. With the ***increment*** of the complexity of the process and plant, new hazards may be generated, and those hazards must be detected early. Someone proposed a method to identify atypical hazards, named the Dynamic Procedure for Atypical Scenarios Identification (DyPASI). DyPASI supports identification and assessment of atypical potential accident scenarios related to substances, equipment and plants based on early warnings or risk notions. Someone proposed a novel model for the failure analysis model based on ***Multilevel*** Flow Modeling (MFM) and HAZOP. Hazard analysis of sub systems is challenging due to their complexity and dynamic component interactions. To perform hazard analysis for such systems, a new hazard analysis technique called SimHAZAN that uses multi-agent modeling and simulation results was proposed research continues to address the limitation of HAZOP and to develop novel methods that can avoid HAZOP's limitations. Someone proposed a computer-aided hazard evaluation method based on ***domain ontology*** called the Scenario Object Model (50M). This methodology can be used to represent the content and structures of the hazard identification process.

2) Dynamic process monitoring for hazard/fault identification. Real time monitoring of process operations and process upsets are of ***paramount*** importance to establish required safety measures. Effective and timely identification of process faults is vital to prevent or control major accidents. The Principal Component Analysis (PCA) method is widely used as a statistical fault detection technique. Harrou et al. proposed a PCA method based fault detection algorithm along with a Generalized Likelihood Ratio (GLR). The advantage of using GLR is that it is able to model the fault detection even in the absence of the process model. The framework to enhance maintenance decisions based on real time information obtained from process monitoring was developed by Elhdad et al. In this method, a combination of real time signals that were triggered during the plant safety shutdown process, ontology and business rules were used to facilitate stakeholders and management making the right maintenance decisions for a ***petroleum*** plant.

Ni et al. proposed a method to predict the location of pipeline leaks using the improved and integrated method of a support vector machine and particle swarm optimization theory (P50-SVM). Small leaks in pipelines can be detected through traditional detection methods which use pressure or a vibration signal. Xu et al. claimed that these traditional methods produced less information from a leakage signal because the high frequency component of a leakage signal weakens rapidly. Therefore, they proposed a novel detection method based on acoustic waves.

A multivariate risk-based fault detection and diagnosis technique was proposed by Zadakbar et al. The proposed technique was capable of eliminating faults that are not significant and providing dynamic risk indication at each sampling stage. Their work shows that use of a Kalman filter combined with the risk assessment method provided robust analysis of false alarms.

(2) Risk Assessment

1) Dynamic risk assessment and management. Dynamic risk assessment has many advantages over traditional static risk assessment. The dynamic changes of the hazardous conditions of highly complex technical and social systems can be modeled using various methods. Further, quantification of uncertainty

helps to improve precision of the risk calculations. The real time updating of risk provides better assessment of risk and thus better management of risk. Adopting dynamic risk assessment and a management model help to predict an abnormal situation and thus better inform decision makers for early actions. This way abnormal events can be prevented before they occur rather than relying on end of the pipe safety measures. This way dynamic risk management helps to enhance the ***inherent*** safety aspect of process operation.

2) Dynamic risk assessment methods. Application of Bayes theorem to update the failure probabilities taking into consideration the dynamic changes of the system has also been discussed to some extent prior to 2013. This section focuses on further development.

Integration of Bayesian Network (BN) with many qualitative and quantitative risk and hazard assessment methodologies enhances accuracy, quantitative power, and reduces the uncertainty. Further, properties of BN such as updating, prediction and forward and backward inference provide significant information for decision makers to make accurate and timely safety critical decisions. The BN facilitates risk quantification even in the case of scarce information and for complex models Pasman and Rogers discussed the application of BN along with Layer of Protection Analysis (LOPA) and explained using a case study. A system with three protection layers was investigated using a discrete BN model and a mixed continuous-discrete BN model. Cai et al. used the Dynamic Bayesian Network to quantitatively assess the impact of human errors on offshore blowout accidents. In this method, a causal relationship was modeled using a pseudo-fault tree and then was converted into BN taking repair of faults into consideration. The important observation of the results was that the human factor failure probability of a barrier that was applied to prevent human errors reached stability when the repair was considered, whereas it increased continuously when the repair action was not considered. State-of-the-art application of BN in FTA for systems for which the Minimal Link Sets (MLSs) and Minimal Cut Sets (MCSs) are known was presented by Bensi et al. Model and parameter uncertainty reduction using Bayesian analysis was discussed by Droguett and Mosleh.

A mapping algorithm to convert traditional bow-tie to BN was proposed by Khakzad. This proposed method differs from other mapping techniques as it demonstrates the fact that probability adapting is more effective than probability updating in dynamic safety analysis. The advantage of probability adapting is that the effects of the generic prior probability reduce since they are updated based on accident precursors or observations Khakzad et al. demonstrated the application of the BN to offshore drilling safety assessment. Khakzad et al. furthered their research and used discrete-time BN to solve the dynamic fault tree without the remedy of Markov chains.

3) Uncertainty handling methods during dynamic risk assessment. During the last two decades fuzzy logic and Bayesian analysis were two common methods used in process safety and risk assessment in order to deal with uncertainties.

Vast possibilities exist with these two methodologies to further extend new model development that would be able to overcome limitations of previously developed methods; hence, results will be closer to reality. Jamshidi proposed a novel method that integrates Relative Risk Score (RRS) methodology and fuzzy logic for pipeline risk assessment. RRS methodology has been identified as one of the most popular pipeline risk assessment techniques developed based on an indexing approach; thus, lack of information and uncertainty were not able to produce realistic information. Integrating ***fuzzy*** logic with the traditional RRS method produces more accurate and realistic results and reflects the real situation. This method was able to take into account the relative importance among the parameters influencing pipeline damages such as third-party damage, corrosion, design, incorrect operation, product hazard, leak volume, dispersion, and receptors.

A fuzzy and evidence based approach along with sensitivity analysis to tackle the uncertainties of input data and model adequacy of the bow-tie model was proposed by Ferdous. This methodology has certain improved features over other methods currently developed. This methodology can ***accommodate*** experts' knowledge to handle uncertainty due to lack of information. The combined approach of fuzzy and evidence based theory addressed the subjective uncertainty, uncertainty due to ignorance and inconsistency associated with experts' knowledge. The model uncertainty was handled through introduction of dependency coefficients. Further, sensitivity analysis was proposed to identify the most significant contributing events to the final end event.

Uncertainty due to ***Common Cause Failure*** (***CCF***) and Diagnostic Coverage (DC), also known as epistemic and aleatory influencing factors, which arise during Safety Instrumented System (SIS) performance analysis, was handled using fuzzy multiphase Markov chains. To incorporate the uncertainty of basic parameters of systems and their impact on SIS performance, fuzzy numbers are used for elementary probabilities in Markov chains. This method illustrated how the imprecision induces changes in the safety integrity level of a particular SIS. The utilization of approximation to estimate the uncertainty information associated with LOPA without using a partial ***derivative*** was discussed by Freeman.

The probability assessment for expert knowledge contains a certain level of uncertainty, known as epistemic uncertainty. There are several methods developed to reduce the epistemic uncertainty that arises during probability assigning. Among them, possibility representation has been identified as an adequate method, especially when informative hard data are not sufficient to perform statistical analysis. Therefore, Flage et al. proposed an integrated probabilistic and possibilistic framework method to analyze the epistemic uncertainty associated with basic events of FTA. The epistemic and objective dependences of basic events and their effects on the top event of the fault tree were analyzed using Frechet bounds and the distribution envelope determination method (Pedroni and Zio, 2013). Results of this study concluded that both types of dependencies significantly affected the top event; however, the epistemic dependencies may have a higher contribution than objective dependences.

(3) Safety Management

A combination of different methods and models is a present trend of developing accident models, which could provide more reliable accident analysis. Wang developed an accident analysis model by combining Human Factor Analysis and Classification System (HFACS) and BN. The causes and prevention measures were proposed using an integrated HFACS-BN model, and then the Best-Fit and Evidential Reasoning (ER) methods were used to rank the proposed safety measures in terms of their cost effectiveness. The blowout accident scenario was modeled using the accident barriers. The safety barriers were proposed based on primary and secondary well control barriers and extra well monitoring barriers. Another five barriers: ignition prevention, escalation prevention, emergency response, blowout control and spill control were proposed to mitigate and control the consequences due to a blowout event. It is also developed an accident occurrence model for the risk analysis of industrial facilities based on the chemical reaction. The model introduced a defensive barrier to prevent a chemical accident that initiates chemical reaction. The uncertainty associated with the barrier was quantified using gamma distribution. Rathnayaka et al. (2013) proposed an accident modeling and risk assessment framework based on accident precursors information. The framework was developed based on SHIPP methodology and applied to deep water drilling operations.

(4) Inherent Safety

Inherent safety/inherently safer design is an emerging concept which generated research and industrial attention during recent years although the concept was established many years ago. Authors described papers published during the last two decades related to inherent safety. In this section, authors highlight further development of inherent safety. Process operations and equipment design considering ISD options at an early design stage could be a better option to achieve higher safety standards and cost benefits. Application of inherently safer design strategies of intensification, attenuation and limitation of effects to model and design the Low Pressure Chemical Vapor Deposition (LPCVD) furnace machine (reactor) was discussed by Chen et al. Application of ISD principles to a laboratory setting where experiments are performed in extremely hazardous conditions was discussed by Theis and Askonas. It is required to know the risk level of the system to successfully apply ISD strategies during the preliminary design stage. Shariff and Zaini proposed a technique which is based on a 2-region risk matrix concept to estimate the risk level at the preliminary design stage. The risk was estimated using a traditional method which is the product of severity and probability and information required to assess the severity was obtained through a process simulator called iCON. Based on risk level, the requirement of inherent safety was decided.

Most of the past developments of inherent safety evaluation indices are hazard-based developments. However, scholars are currently focusing on both hazard reduction and likelihood reduction through application of ISD principles. As a result risk based indices are proposed. Rusli proposed a framework called Quantitative Index for ***Inherently Safer Design*** (QI2SD) that evaluates the inherent safety level of the system in terms of element of risk rather than only considering hazard reduction. In addition to quantitative evaluation, it is capable of evaluating the hazard conflicts and trade-off which may arise during the application of ISD strategies and it facilitates the ranking of ISD alternatives for decision making.

3. Future Direction

The systematic explanation of the methods and models developed, starting from origin to current research, provides a natural guide to future direction of research. It is clear that the current research trend has been in the area of inherent safety, dynamic and operational risk assessment, incorporation of human and organizational factors into risk assessment and integration of a safety protection layer (safety instrumented system) into risk assessment.

Transition from traditional quantitative risk assessment to Dynamic Quantitative Risk Assessment (DQRA) is a natural evolution. DQRA enables implementation of inherent safety principles, features most desired in hazardous processes.

New Words and Expressions

simultaneously [ˌsɪməl'teɪniəsli] *adv.* 同时
increment ['ɪŋkrɪm(ə)nt] *n.* 增量，定期的加薪，增加
multilevel [ˌmʌltɪ'levəl] *adj.* 多级的，多层的
domain ontology 领域本体
paramount ['pærəmaʊnt] *adj.* 至为重要的，首要的，至高无上的，至尊的
petroleum [pə'trəʊlɪəm] *n.* 石油，原油

inherent [ɪnˈhɪər (ə) nt] *adj.* 固有的，内在的

fuzzy [ˈfʌzi] *adj.* 覆有绒毛的，模糊的，失真的，头脑糊涂的

accommodate [əˈkɒmədeɪt] *v.* 容纳，顺应，提供住宿

Common Cause Failure (CCF) 共因失效

derivative [dɪˈrɪvətɪv] *n.* 衍生物，派生词，衍生字，派生物

Inherently Safer Design 本质较安全设计

Unit Thirteen

13

Workers' Compensation Insurance and Occupational Injuries

Occupational accidents not only threaten employees' lives and damage employers' human ***capital***, but also increase the social costs of a country. In 1998, the global number of fatal injuries was estimated at 350,000, or 970 deaths per day. The number of non-fatal occupational injuries with three or more days' absence from work was also 264 million, which is more than 700,000 injured workers per day. In addition to the numbers of accidents, the cost of the accidents is also huge. The International Labor Organization (ILO) estimated that the total cost of occupational accidents and work-related diseases are 4% of gross national product of a given country.

To reduce these damages, almost all industrialized countries have introduced laws and regulations for the prevention of occupational accidents and work-related diseases. Among others, some form of government-mandated insurance guarantees compensation for injuries or diseases arising ***out of employment*** in virtually all nations (The US Social Security Administration has compiled a description of social insurance mechanisms in 112 countries. These descriptions are contained in a series of publications entitled Social Security Programmes through out the world. Separate reports are published for four different regions of the world: Africa, the Americas, Asia and the Pacific, and Europe). In reality, however, countries vary greatly with respect to how they organize workers' ***compensation*** systems in terms of the sources of funds, the mechanisms used, and the allocation of system costs among employers and others. Thus, it is easily predictable that these different approaches may have significant implications for system performance, including employers' and workers' incentives to promote workplace safety and hence impact the actual the occurrence rates of occupational injuries and diseases.

In this study we ***investigate*** the consequences of various terms of conditions in ***workers' compensation insurance***, e.g., types of systems used, employer's payment mechanisms, and coverage of compensation for the injured worker, on the occurrence of occupational injury and disease. One distinctive feature of the study is that we address this issue in the context of cross-country ***empirical*** analysis. Previously many studies have performed similar research within a specific industry groups (in a specific kinds) of occupational injury. It is certain that those kinds of studies shed more light on the status of the small-scale picture of the relationship between workers' compensation insurance and occurrence of occupational injuries and diseases. However, when policy-related issues are involved, comparisons and results with a rather widely designed setting is more appropriate since it can more directly reveal the underlying economic effects of a workers' compensation insurance design.

One of the critical issues in the cross-country empirical analysis is how to control the possible heterogeneity problems among the countries. Each country has a different definition in clarifying the fatal and non-fatal occupational injuries and diseases, and each adopts different reporting systems in constructing statistical databases. This means that the data may have serious measurement error problems and different degrees of under (over) estimation problems. Thus, it is important for the study to control the ***heterogeneity*** issues. In this study we try to control these heterogeneities by using as many data sources as possible and identifying the data sources with various specification methods. A fixed effect model with ***panel*** data is also employed to control unobserved time invariant country specific fixed effects.

After controlling for the various country fixed effects and several relevant aggregate variables, we found much important empirical evidence that the workers' compensation system is significantly related with the occurrence of occupational injuries and diseases. First, the regression result suggests that a private insurance system would be more efficient than a public (social) insurance system in that it lowers the occurrence of occupational injuries and diseases. The result implies that the efficiency and cost-reduction rationale for a private insurance system are stronger than the equity-based reasoning for public insurance for the reduction of occupational accidents. Second, risk-based pricing for the payment from an employer is correlated with higher occupational injuries and diseases. This result seems to be surprising since it contradicts the conventional wisdom that a risk-based approach internalizes the hazards of occupational accidents, resulting in the reduction of the occurrence of occupational injuries and diseases. Third, the result shows that the wider the coverage of injured workers is, the less frequent workplace accidents are. The result implies that the worker-comfort effect may surpass the moral hazard effects in the workplace.

The analysis below investigates the factors affecting occupational injuries and diseases through cross-country ***regression*** analysis. We, in particular, focus on the relationship between the workers' compensation system and the occurrence of occupational injuries and diseases. The main purpose of the study is to examine how the workers' compensation system, meaning aspects such as the types of systems, the payment mechanisms, and coverage of compensation, affect the occurrence of occupational injuries and diseases.

Although compensation for injury or disease arising out of employment is guaranteed in virtually all nations, the degree of compensation varies greatly ***with respect to*** how each country operates and maintains workers' compensation systems in terms of the sources of funds, the system used, the allocation of system costs among employers and others, and the actual coverage for injured workers. These different approaches inevitably generate significant implications for system performance, including employers' and workers' incentives to promote workplace safety.

One of the most debated issues in the design of workers' compensation insurance is whether private insurers should be permitted to operate in the traditionally-held social (public) insurance system. Naturally the first aspect of the workers' compensation insurance in Organization for Economic Cooperation and Development (OECD) countries is the difference in the "types of systems". We decompose the type of systems into two primary categories: social (public) insurance and private insurance (Although countries such as Thailand and Nigeria have "employer liability" system, we exclude them since no OECD countries adopted this system).

Traditionally, proponents for government monopolies (social insurance) address the "market failure" problem that could appear in workers' compensation insurance. They argue that private companies may raise the workers' compensation rates of individual employers in the search for profits. The reason behind this is that private insurers would lead to "adverse selection", insuring only low-risk employers, and the national agency acting as society's insurer of last resort will have more burden than ever. They also argue that public insurance may be necessary to capture economies of scale that may exist in the insurance industry (Besides the economic efficiency issue mentioned above, some proponents argue that workers' compensation insurance should be developed as a "pure social insurance" and a fundamental right or entitlement of citizenship. In this view, the unavoidable cross-subsidy in which the low-risk groups subsidize the high-risk groups is judged to be good for the equity reason of ***redistribution*** and could be achieved in the public insurance system).

On the other hand, proponents for a private insurance system argue that the entrance of insurance firms into the market would induce efficiency and lower the costs of any given level of comparable insurance services. The argument is that the government monopoly, not being subject to the same competitive pressures that the marketplace forces on private firms, would not provide the appropriate combination of services and not administrate their system more efficiently than would a private insurer. This means that the public insurance system would not adopt more efficient cost-saving technologies, while the private sector would.

We believe that the two systems will have different effects on the incentives of employers and workers. All other things being equal, the economic incentive for a private company to reduce the costs of any insurance services will affect the occurrence of occupational injuries and diseases (Some researchers, such as Klein and Krohm see that the choice of public or private delivery mechanisms appears to be more a matter of a nation's preference). Hence, although various rationales of proponents for both public and private insurance can be justified as possible explanations, we need an empirical investigation to see whether a particular system is helpful for the actual reduction of occurrences of occupational injuries and diseases.

The second aspect we focus on is the difference in the employers' funding mechanisms. With respect to how employers are required to fund workers' compensation costs, we distinguish fixed flat rates (tied to pay roll) from risk-based rates that vary among employers according to a system for assessing their relative risk level (In addition to the two categories, there are non-OECD countries such as Pakistan, Bangladesh, and Nigeria where employers directly reimburse their workers' compensation costs).

The economic incentives in the form of risk-based pricing or graduated premiums strongly promote the effort of employers for the prevention of accidents and the avoidance of occupational injuries and diseases in the workplace. The key feature of risk-based pricing is that employers with the same level of risk should pay the same cost according to the inherent hazards of a class of employment or the actual loss experience of an employer. This means that the employers hiring workers for the higher risk jobs will pay a higher insurance price than the employers hiring low-risk workers. The rationale for this approach is that the risk-based contribution or premium is not only equitable but it is also efficient in the sense that it internalizes their risk and the associated costs. It is often asserted that the risk-based pricing of insurance has strong theoretical appeal to economists and intuitive acceptance by risk managers and insurance professionals.

The third aspect is related to the difference in the compensation coverage for injured workers. With respect to the coverage, two variables, the maximum period of absence day and the extent of income substitution, are considered as proxy variables representing the compensation coverage for injured workers. The absence day of the sample OECD countries varies from six months to unlimited periods, and the level of income substitution ranges from 50 percent to 100 percent. We categorize these variables using a 5-unit Likert Scale and examine the relationship between the coverage of compensation and occurrence of occupational injuries and diseases.

It is interesting to note that virtually all studies of claim usage in workers' compensation insurance find that an increase in indemnity benefits increases claim frequency. For the reasons behind these results, Butler and Worrall emphasized two different types of moral hazards. One is the "risk bearing" moral hazard in which higher benefits induce workers to take more ex ante job risks given a higher level of ex post injury compensation. The other is the "claims reporting" moral hazard in which higher benefits have no effect on actual injuries (risk is unchanged) but does induce more claim filings.

While it is true that increasing coverage leads to more risk bearing and claims report of behaviors on the part of workers, we think there is another dimension to be considered. That is, the so-called "worker comforting" effect in which higher benefits induce workers to have more secure and comfortable psychological status than before, and thus to work in a safer way for given work related risks. We believe that these two different dimensions-moral hazard and worker comforting effects should be checked empirically with cross-country data.

Controlling for heterogeneity and multi-collinearity problems with a cross-country fixed effect model, the regression results provide empirical evidence that the compensation system plays an important role in explaining the occurrence of occupational injuries and diseases among OECD countries. A private insurance system, fixed flat rate employers' funding mechanism, and higher compensation coverage scheme are significantly and positively correlated with lower levels of occupational accidents compared to the public insurance system, risk-based funding system, and lower compensation coverage scheme. These results indicate that employers under private insurance systems promote their efforts more to reduce occupational accidents and the associated insurance costs by internalizing the risk of the work place. Higher coverage is also associated with lower occurrence of occupational injuries and diseases, which indicates that the "worker comfort effect" is greater than the "moral hazard effect". These results mean that the insurance compensation system plays an important role in determining the occurrence of occupational injuries and diseases and the policy makers need to consider the effect of the insurance systems in order to reduce the occurrence of occupational accidents.

Another important finding is that the factors representing industry structure (such as the ratio of construction industry and the ratio of manufacturing sector over the economy) have stronger effects on the occurrence of fatal accidents, while the population structure or gender difference is more ***attributable*** to the occurrence of non-fatal accidents. These empirical results suggest that the policy makers need to be aware of the relative effectiveness of policy instruments and set policy priorities when they focus on a specific type of occupational accident.

The critical econometric issue of the cross-country study is how to control the possible heterogeneity problem among the countries. In this study, we use as much data as possible to identify the heterogeneities as well as panel data fixed effects models to control them. The cross-country differences in the source of gathered data on occupational injuries and diseases could also be identified with the ILO LA-BORSTA database. However, we need to mention that, if unobserved characteristics of countries change over time in ways that affect occupational accidents differentially, these effects are not captured by the fixed effects model, and the relationship between workers' compensation insurance terms and accidents frequency may not be correctly identified. Even though it is reasonable for us to assume that several data differences definitions of occupational accidents, lack of data, and underestimation practices—may have constant effects within a country for the given time periods (1990–2008), we need to check the validity of the assumption. A more detailed treatment of these issues awaits further work.

New Words and Expressions

capital [ˈkæpɪt(ə)l] *n.* 资本，资金，首都，大写字母

out of employment 失业

compensation [ˌkɒmpenˈseɪʃ(ə)n] *n.* 抵销，补偿

investigate [ɪnˈvestɪgeɪt] *v.* 调查，研究

workers' compensation insurance 工伤补偿保险

empirical [emˈpɪrɪk(ə)l] *adj.* 经验主义的，完全根据经验的，实证的

heterogeneity [ˌhetərəʊdʒɪˈniːəti] *n.* [生物] 异质性，[化学] 不均匀性

panel [ˈpæn(ə)l] *n.* 镶板，仪表盘，钣金，(衣服上的) 镶条

regression [rɪˈgreʃ(ə)n] *n.* 回归，退化，倒退

with respect to 关于，至于

redistribution [ˌriːdɪstrɪˈbjuːʃən] *n.* [经] 重新分配

attributable [əˈtrɪbjʊtəbl] *adj.* 可归因于，可能由于

Exercises

1. Explain the characteristics of occupational accidents.
2. What do they investigate in this study about workers' compensation insurance?
3. What are the critical issues in the cross-country empirical analysis?
4. Simply describe the two different types of moral hazards that Butler and Worrall emphasized.
5. How to control the possible heterogeneity problem among the countries?

科技英语翻译技巧——否定句的翻译

在科技英语中，表示否定意义的词语及其用法与汉语中的词语及其用法并不完全相同。否定是英语里使用很广泛的表达方式，有全部否定、部分否定、双重否定、意义上的肯定、含蓄否定和转移否定。

1. 全部否定

英语中的否定，按照否定范围，可分为全部否定和部分否定：全部否定是指否定整个句子的全部意思；部分否定是对句子的一部分进行否定。全部否定常用的否定词有：no、never、none、nor、neither、nobody、nothing、nowhere 等。在翻译时，不论这些词在句中是状语、定语、主语或宾语，都可直接翻译成否定句。

Liquids are different from solids in that liquids have *no* definite shape.

液体不同于固体，因为液体没有一定的形状。

Carbon dioxide does not burn, *neither* does it support burning.

二氧化碳既不自燃，也不助燃。

Never is aluminium found free in nature.

铝在自然界从不以游离状态存在。

2. 部分否定

英语的部分否定是由 all、every、each、both、always、much、many、often 等词与 not 构成的。not 的位置灵活，可在这些词之前，也可位于其中，相当于汉语中的“不是所有都”“不是每个都”“不是两个都”“不总是”。当否定词 not 和谓语构成否定句时，形式上是全部否定，实际上是部分否定。

Every electric motor here is *not* new.

这里的电动机不全是新的。

Notice that *not all* mechanical energy is kinetic energy.

要注意，并非所有的机械能都是动能。

In a thermal power plant, *all* the chemical energy of the fuel is *not* converted into heat.

在火力发电厂里，燃料的化学能并未全部转化成热能。

3. 双重否定

双重否定是两个都表示否定意义的词出现在同一个句子中形成的，即否定之否定。英语中的双重否定形式通常由否定词 no、not、never 等表示否定意义的词构成，如 no... without, without... not, never... without, not/none... the less, not/never... unless, no less... than, not... until, not... any the less, impossible... without 等。

It is *impossible* for electricity to be converted into certain energy *without* something lost.

电转换成某种能量时不可能不产生损耗。

Sodium is *never* found uncombined in nature.

自然界中从未发现不处于化合状态的钠。

4. 意义上的肯定

英语中有些否定词与其他词构成否定意义的词或词组，因而在形式上是否定的，但实际意义上是肯定的。翻译这类句子时，要结合上下文，翻译成一般形式的肯定句或者更加强势的肯定句。常见的形式有 nothing like“没有什么比得上……”, no/not/nothing + more + than“不过”, no/not/nothing + less + than“多达”等。

“I had seen fossils before, but *nothing like* this,” he says.

他说：“我见过化石，但从来没有见过这样的化石。”

Highly processed foods have these flavour enhancers which are *nothing more than* carcinogenic chemicals with no natural flavours of their own.

精加工食品都含有这些增香剂，这些无非是一些致癌物质，本身并没有什么自然的香味。

The lack of verbal distinction between different types of “dirty” has not made communicating the idea of recycling easy, but now the tribe has *no less than* five different sacks to choose from.

缺乏语言来表达各种不同的“脏东西”使向村民们传递回收的概念不太容易，但是现在部族里有了不少于五种分类的垃圾袋可选用。

5. 含蓄否定

英语中有大量的词语和句子，尽管在形式上没有否定词或者否定结构，但意义上是否定的，这种称为含蓄否定。

表达含蓄否定的动词和动词短语：avoid，deny，exclude，fail，ignore，lack，neglect，overlook。

表达含蓄否定的名词：absence，deficiency，exclusion，failure，loss，neglect，refusal。

表达含蓄否定的形容词：short of，safe from，little，last，free from，devoid of。

表达含蓄否定的副词：hardly，rarely，scarcely，seldom，vainly。

表达含蓄否定的介词及介词短语：above，below，beneath，beside，beyond，but for，instead of，behind，in place of，in vain，in defect of。

表达含蓄否定的连词：before，but，but that，rather than。

The robot *seldom* if ever tires or losses interest in its task.

机器人对于所承担的工作几乎不会感到厌倦或失去兴趣。

A vacuum, which is the *absence* of matter, cannot transmit sound.

真空中没有物质，不能传播声音。

6. 转移否定

英语中的否定词往往出现在否定部分的前面，对后面进行否定，但有时否定词所否定的并不是紧随其后的部分，而是后面的某部分，这就构成了否定转移。英语中的否定分为一般否定转移和特殊否定转移两种形式。前者否定的是谓语部分，后者否定的是谓语以外的其他部分。

The motor did *not stop because* the electricity was off.

电动机停止运转，并非因为电源切断。

Liquids, except for liquid metal as mercury, *are not considered to be good conductors of heat.*

液体除了液体金属水银外，都被认为是不良导热体。

The planets *do not go* around the sun at a uniform speed.

各行星不是以相同的速度绕太阳运转的。

Reading Material

Occupational Injury Insurance's Influence on the Workplace

Worker's compensation provides health care and partial wage loss replacement to workers injured on the job. It is more ***complex*** in its scope and impact on workers than other social insurance both because it is a system of diverse state-based laws funded through private, public, and self-insuring entities, and because it has significant overlap with health insurance, unemployment insurance, and other employer-provided benefits. While research indicates strong incentive responses to the structure of indemnity benefits and medical reimbursements, future research will benefit from employing a worker-centric (rather than a program-centric) orientation using integrated databases that ultimately link workplace productivity to program characteristics.

If John Q. Public is injured on the job or acquires a job-related disease, all his medical expenses and possibly some of his lost wages, are covered in the USA and Canada by Workers' Compensation (WC). WC is a system of no-fault laws implemented, state by state, and province by province, about 100 years ago. While the system is fragmented, it is large: In 2008, $ 58.3 billion of WC benefits were paid out to workers in the USA, over half going for health care expenditures and slightly less than half for lost work time pay. The cost to the US firms of the system in 2009 was $ 74 billion.

While we describe some ***institutional*** details of the US system, Canada and many developed nations have similar systems for workplace injuries. Because each state dictates how wages are to be replaced (with "indemnity benefits"), who can initially direct the health care of the injured worker, and what mechanisms the firm can employ to provide for this insurance coverage, this is considered "social insurance". Though each state has its own WC system (and a Federal law to cover Federal employees), the institutional characteristics of WC are broadly consistent across states and similar to many systems outside the United States. Hence, the medical, redistributive, and incentive issues applicable to the WC system in the USA carry over to other countries.

1. Benefits

With respect to employee benefits under WC, all state laws require nearly 100% coverage of medical expenses and some minimum cash benefits related to lost earnings for those out of work longer than the state-specific waiting period (2–7 days, with modal waiting periods of 3 and 7 days). Though WC benefits are more generous than unemployment insurance benefits, the structure of reimbursements is similar. John Q. Public receives no lost workday compensation (i. e., no indemnity compensation), unless his claim exceeds the state's waiting period.

Once past this waiting period, he is classified as a temporary total injury and receives approximately 67% of usual weekly wages as a benefit, subject to a state-wide weekly maximum payment, often set equal to the average wage for the state. After the injured worker's medical condition has stabilized (called the point of "maximal medical improvement"), his injuries are evaluated for the permanency of the condition. Depending upon the state, he may be paid a lump-sum for things such as an amputated finger (scheduled benefits) or receive a weekly payment (benefits not stipulated for specific injury types by state legislation, such as low back pain, are "unscheduled"). In some states, there are time limits on the duration that a worker can receive benefits, though the most severe injuries generally have no time limits.

If the injury lasts longer than the retroactive period, the associated benefits withheld during the initial waiting period are reimbursed to the worker. Medical costs associated with a workplace injury are compensated on a ***fee-for-service*** basis, employing the "usual and customary" fees for the local health care market. Whether chiropractic care is reimbursed depends on state statues regulating WC claims. WC policies are occurrence policies, so that if the injury happens when the insurance is in force, all associated payments for lost time or medical costs are incurred immediately and continue even if the worker doesn't return to work or the firm changes insurance coverage. The laws also provide for rehabilitation services and the payment of income benefits to dependents in the case of a workplace fatality.

2. Insuring Workplace Risk

Except for ***domestic*** servants, agricultural workers in some states, and some very small employers, WC laws make employers liable for all medical expenses and a portion of lost wages for their injured workers. The employer pays these benefits for injuries arising "out of and in the course of employment", regardless of who is at fault, except under certain exclusions, such as when the worker is under the influence of alcohol or drugs. The liability imposed on the employer is ***exclusive***: The injured worker cannot sue for additional compensation nor is there compensation for pain and suffering.

Some large firms and groups of moderately sized firms are able to self-insure their liabilities. Generally, however, employers are required to purchase insurance to cover their potential liabilities. In four states—North Dakota, Ohio, Washington, and Wyoming—the insurance for those not self-insuring is provided by monopoly state funds. The remaining 46 states provide that insurance be purchased from private insurance carriers or, in 20 states, from a state fund that competes with the private carriers.

The procedures for calculating insurance premiums are similar across states. To calculate the firm's insurance premium, the firm's workers are placed into one or more of approximately 600 industrial-occupational classifications. On the basis of these classifications, the firm is assigned "manual rates" which are ***premium*** rates reflecting the average losses found in each classification. "Manual premiums" are then calculated by multiplying the manual rate by the payroll of workers in the classification. These manual premiums are summed for all employee classifications to arrive at the manual premium for the firm.

The actual premiums paid by the smallest firms are these manual premiums. The vast majority of all employers in the USA pay unmodified manual premiums, though these employers have relatively few employees and only account for a small fraction of all employment covered by WC.

If the manual premium exceeds a given amount, then premiums are "experience-rated". In this case, the manual premium is modified to reflect the firm's own past injury loss experience: If the firm experiences fewer than expected WC claims in one of their classifications, the manual rate is adjusted downward to reflect the firm's better than average experience. If the firm is worse than average, rates ***go up***. In the USA, the premium of an experience-rated firm is a weighted average of the manual premium and the firm's actual loss experience, where the weight placed on actual loss experience grows with firm size. That is, the extent to which a firm's actual premium reflects its own injury losses depends on the size of the firm. Small firms pay experience-rated premiums largely reflecting the manual premiums. In contrast, large experience-rated firms pay premiums largely reflecting their own loss experience rather than the average loss experience. The weight placed on the firm's own loss experience is termed the "degree of experience-rating". The greater the degree of experience rating, the more strongly the firm's incentives are tied to loss prevention, including incentives to invest in workplace safety and claims management. Firms that self-insure bear all of the costs of WC benefits directly, resembling full experience-rating. However, simulations by Victor have shown that fully experience-rated premiums can provide stronger incentives for safety than self-insurance.

3. How Complete Is Workers' Compensation Coverage?

As indicated above, WC benefits and experience rating have been found to generate moral hazard responses by firms and workers. In the absence of moral hazard responses in safety and reporting behavior induced by the insurance coverage, full coverage is ***optimal*** with risk averse workers. With moral hazard responses, optimal insurance is less than 100% wage replacement—hence, the standard cost sharing WC insurance mechanisms of partial wage replacement and waiting periods. Moreover, hard-to-monitor claims such as low back claims seem to exhibit more moral hazard response than other claims (Dionne and St-Michel, 1991), so one Pareto—improving adjustment to the benefit structure (that could be cost neutral within the WC system) would raise the replacement rate of easy to monitor injuries (such as lacerations and broken bones) while lowering the replacement rate of hard-to-monitor injuries (such as low back claims). Finding such patterns of benefit response in their data, Johnson suggest such a change in the WC benefits structure.

Bronchetti and McInerney estimate that WC benefits significantly smooth consumption lost to workplace injuries, even in the presence of potentially low take-up rates, and given the moral hazard responses historically measured (given in the previous two sections above), finds that WC benefits may be slighter higher than is socially optimal (weighing the importance of consumption smoothing against moral hazard response).

In conjunction with these considerations of the optimal structure of WC payments by moral hazard potential, there is a small literature on how fully WC benefits replace lost wages of workers injured on the job. These estimates have to be made with care, particularly for inter-temporal comparisons. For example, some workers with long-term back pain may not be ***eligible*** for WC benefits if the back pain is not related to their current job. Modern sedentary life may change the incidence of back pain in uncertain directions. Moreover, firms are increasingly using restricted work for injured workers as benefits increase and waiting periods shorten, changing the estimated dynamic of days away from work and benefits.

Partial wage replacement mechanisms that guard against moral hazard response also bias the estimates of lost earnings. Waiting periods ***guarantee*** that most minor workplace injuries will end up being medical only claims, and hence, there will be no wage replacement even in the absence of work-origin or "take-up" rate problems. Using countrywide data on claims, Appel and Borba report that 81.2% of all claims are medical only claims (using an ultimate report basis). This high fraction of medical only claims is also apparent in more recent data (New Mexico Workers Compensation Research Department reports that 76% of their claims were medical only in 2010). Even so, evidence suggests that not all claim-eligible workers get WC benefits, both because the take-up rate for claims is less than 100% even past the waiting period and because some claims are denied by the employer/insurer.

Most studies of WC take-up rates compare administrative WC data with some alternative measure of injured workers. For their alternative measure of work injuries, Biddle and Roberts employ a sample of workers with shoulder, back, wrist, or hand sprains/strains whose injuries were deemed work related by physicians. The advantage of these injuries is that they cover a relatively large fraction of WC claims and so are fairly representative of typical claims; the disadvantage is the specific work origin of a soft tissue injury may be difficult to determine, even by a worker if they have more than one job for any length of time. In comparing their sample of injured workers with Michigan administrative WC data, they find that up to 60% of their injured workers never filed a WC claims, primarily because workers self-report that they did not think the condition was sufficiently serious. Even among those with more than one week of ***lost work*** (beyond the Michigan waiting period), 40% did not file for wage-loss benefits. An earlier analysis suggests that the low take-up rate is not just a strain/sprains problem as they find acute conditions just as likely to be under-reported as ***chronic*** conditions.

As WC insurance is purchased by the employer and WC claims are filed by workers, Coase theorem outcomes are less likely to be obtained between firms and workers with respect to claiming behavior and wages (expect perhaps between large, self-insuring firms and long-tenured skilled employees). Hence, it is feasible that higher denial rates could lower claims filing, particularly, for hard to monitor injuries like low-back pain. Biddle (2001) finds such a result in an Oregon sample.

Boden and Galizzi find that evidence that women receive lower loss-wage reimburse-menu than men, which may reflect ***discriminatory*** differences in injury benefits. Using a sample of injured men and women from Wisconsin in 1989 – 1990, they find significant differences in loss-wage reimbursements 3 years after the injury, even after observable labor supply factors are taken into account.

Clearly there are a lot of unresolved problems with estimating wage-loss in WC, and future research—perhaps along the lines of Bronchetti and McInerney (2012)—will be welcome in this literature.

New Words and Expressions

complex [ˈkɒmpleks] *adj.* 复杂的，难懂的，费解的，复合的
institutional [ˌɪnstɪˈtjuːʃ(ə)n(ə)l] *adj.* 机构的，慈善机构的
fee-for-service 医疗费
domestic [dəˈmestɪk] *adj.* 本国的，国内的，家用的，家庭的
exclusive [ɪkˈskluːsɪv] *adj.* 独有的，排外的
premium [ˈpriːmɪəm] *n.* 保险费，额外费用，奖金，津贴
go up 增长，上升，被兴建起来
optimal [ˈɒptɪm(ə)l] *adj.* 最适宜的，最理想的，最好的
eligible [ˈelɪdʒɪb(ə)l] *adj.* 有资格的，合格的，具备条件的
guarantee [gær(ə)nˈtiː] *v.* 保证，保障，担保，确保
lost work 无效功
chronic [ˈkrɒnɪk] *adj.* 长期的，慢性的，难以治愈的，长期患病的
discriminatory [dɪˈskrɪmɪnət(ə)ri] *adj.* 区别对待的，不公正的，歧视的

Unit Fourteen

14

The Current Situation and Development Countermeasure for China Special Equipment Safety Supervision

Special equipment, which is widely used in industrial production and people's life, poses potential danger. Some special equipment work under high temperature and high pressure, some contain flammable, explosive, toxic, or harmful medium, and some run at high altitude or high speed. In the event of an accident, it is prone to cause fire or explosion, casualty, environmental pollution, economic loss and has very bad social influence. Considering this, countries around the world all attach great importance to its safety and gradually establish a scientific and standardized system to ***supervise*** special equipment. As countries around the world differ in development history, cultural background, political system, economic level, and so on, the supervision bodies and supervision measures are different from each other. Some countries put the abovementioned equipment under the responsibility of one department for unified supervision and administration, in some countries, the special equipment are scattered and administrated in several departments of the government (Sun, 2013). In China, special equipment is under unified supervision and administration by specialized government bodies.

1. Development History and Current Situation

(1) Basic Information of Special Equipment

1) Definition of special equipment. Special equipment has not yet formed a unified concept or reached broad consensus internationally. In China, it is a term referring to boilers, pressure vessels (including gas cylinders), pressure pipelines, elevators, lifting appliances, passenger ropeways, large amusement devices, non-road vehicles, and other equipment which is ***stipulated*** by laws or administrative regulations to be subject to the *Special Equipment Safety Law of the P. R. China*. Special equipment has potential danger to safety of human lives and properties. Among special equipment, boilers, pressure vessels (including gas cylinders), and pressure pipelines are collectively referred to as pressure equipment; elevators, lifting

appliances, passenger ropeways, large amusement devices, and non-road vehicles are classified as electromechanical special equipment. Special equipment usually has the following characteristics: They are sealed, pressure-bearing, composed of mechanical components, and usually they are key components of a continuously operating system (Zhong, Zhang, Luo, & You, 2009). China compiles special equipment directory for its supervision (AQSIQ [General Administration of Quality Supervision, Inspection and Quarantine] Notice on the Revision of the Special Equipment Directory [2014] No. 114).

2) Data and safety situation of special equipment. With the fast growth of economy and the development of technology, China has become the country with the biggest growth of in-service special equipment in the world. Over the past decade, its annual increase rate is over 10%, as shown in Table 14-1.

Table 14-1 2005 – 2015 Quantity of Special Equipment

	Boiler (10k)	Pressure Vessel (10k)	Elevator (10k)	Lifting Appliance (10k)	Passenger Ropeway	Large Amusement Device (10k)	Non-road Vehicles (10k)	Gas Cylinder (10k)	Total (10k)
2005	55.38	152.30	65.18	71.03	821	1.59	28.73	12, 981	376.30
2006	54.30	158.80	77.14	82.36	836	1.40	27.21	13, 086	403.73
2007	53.41	169.71	91.73	95.79	814	1.44	28.58	13, 210	443.32
2008	57.82	192.72	115.31	188.28	793	1.47	32.56	13, 244	521.11
2009	59.52	214.32	136.99	135.27	850	1.56	34.81	13, 239	582.56
2010	60.74	233.60	162.86	150.00	860	1.58	38.79	14, 073	647.65
2011	62.03	251.54	201.06	171.74	863	1.64	41.03	13, 564	729.15
2012	63.53	271.82	245.33	190.94	845	1.67	48.29	13, 881	821.67
2013	64.13	301.12	300.93	213.50	873	1.79	55.36	14, 387	936.91
2014	63.89	322.79	359.85	226.26	925	1.92	61.66	14, 250	1, 036.46
2015	57392	340.66	425.96	210.44	985	2.04	63.02	13, 698	1, 100.13

By the end of 2015, the total number of special equipment production units (including design, manufacture, installation, alteration, repair, and gas filling) is 62,706 nationwide, a complete industrial chain has been formed from design, manufacture, inspection, installation, modification to repair, and the annual output is over 1.3 trillion *yuan*. Statistics show that China's special equipment import goes down each passing year; on the other hand, the export increases year by year. China has become the largest producer of heavy pressure vessels such as supercritical power plant boilers and hydrogenation reactors. Its maximum hoisting capacity of bridge crane has reached 22,000 tons, ranking the first in the world. China is the world's largest producer and exporter of non-refillable steel welded cylinders. China also ranks the first in the world for elevator in the aspects of registered installations, annual output, and annual growth rate. Of the annual newly built elevators, the elevator of China made accounted for 70%, in 2014, the registered elevator installations of China accounted for 27% of global share (Song, 2013; Zhang & Zhao, 2015).

By the joint effort of end users, supervision bodies, inspection agencies, research institutions, and the public, the special equipment safety accident rate of China decreases year by year. Special equipment safety situation continues to improve, especially in recent years, it maintains a steady decline trend as shown in Figures 14-1 and 14-2. On the global scale, there is still a large gap on the overall special equipment safety condition between China and industrialized countries, but in regards to some special equipment, the accident rate and mortality of every ten thousand units of equipment are equal to or better than that of developed countries.

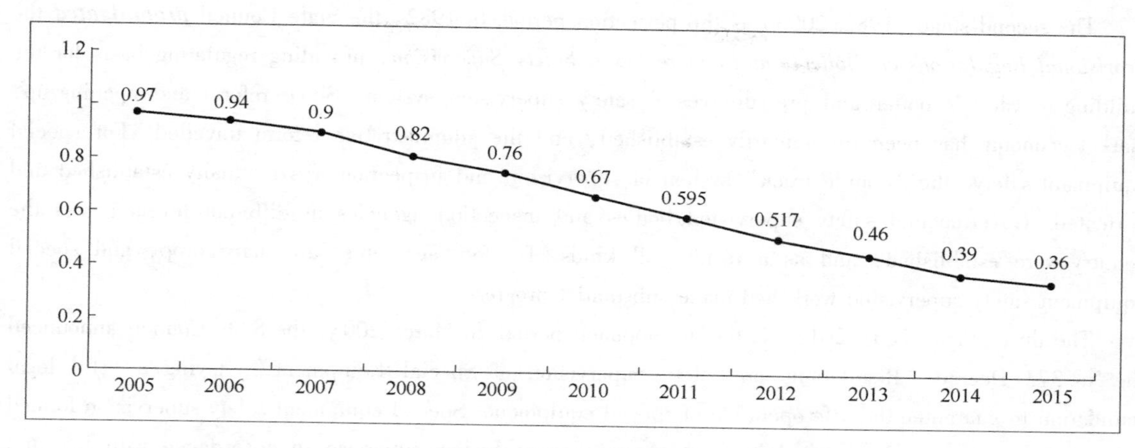

Figure 14-1 2005–2015 China Special Equipment Mortality Rate (People/10k)

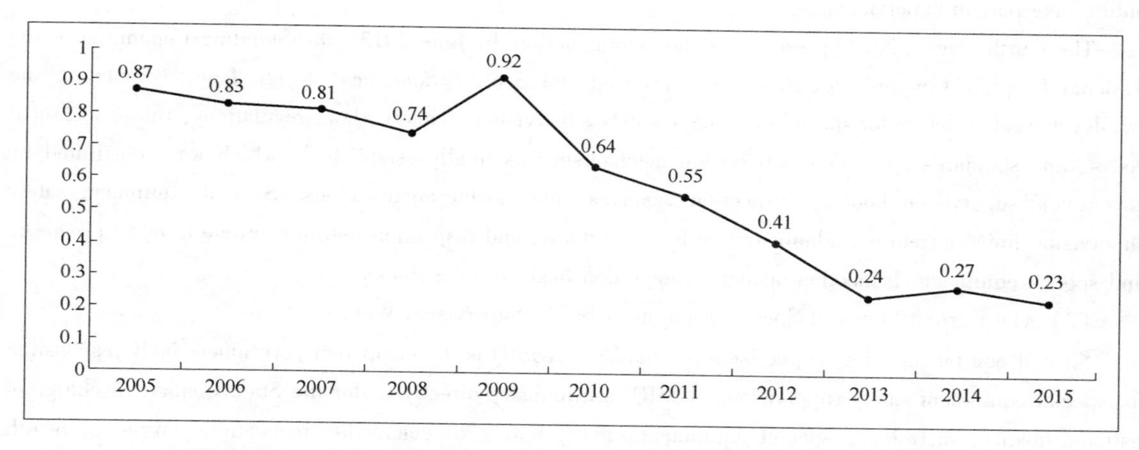

Figure 14-2 2005–2015 China Special Equipment Accident Rate (Case/10k)

(2) Development History of Special Equipment Safety Supervision

China's special equipment safety supervision was gradually developed along with major and extraordinarily serious accidents. On April 25, 1955, the state-run Tianjin First Cotton Mill's boiler exploded, eight people were killed and 69 were wounded. In July 1955, referencing the safety supervision pattern of the Soviet Union, the State Council approved to set up the Boiler Safety Inspection Bureau under the Ministry of Labor to supervise boiler, pressure vessel, hoisting machinery, and other special equipment. Over 60 years of development, China's special equipment safety supervision continues to improve, and according to studies of government bodies, experts, and scholars (Pen, 2009; Nie, 2010; Shi, 2009), it can be summarized into four stages.

The first stage (1955–1982) is the exploration period. Influenced by historical conditions, it had not formed a coherent, systematic safety supervision system. During this period, the safety supervision body experienced twice removal and merger, which severely affected the safety supervision work. After 10 years of "Cultural Revolution", safety accidents were frequent. In 1979, every ten thousand units of boiler and pressure vessel accident rate was 7.9, and the safety situation was severe.

The second stage (1982–2003) is the perfection period. In 1982, the State Council ***promulgated*** the *Provisional Regulations on Boiler and Pressure Vessel Safety Supervision*, providing regulating basis for the building of China's boiler and pressure vessel safety supervision system. Since reform and opening-up, market economy has been preliminarily established, and the administrative reform unveiled. For special equipment safety, the "double track" system of supervision and inspection was gradually established and perfected. Governmental safety supervision bodies and inspection agencies at different levels across the country were established, and as a result, all kinds of safety accidents saw sharp drop, and special equipment safety supervision work had made substantial progress.

The third stage (2003–2013) is the development period. In March 2003, the State Council announced the No. 373 Decree "Regulations on Safety Supervision of Special Equipment", laying a solid legal foundation to guarantee the safe operation of special equipment. Special equipment safety supervision formed the working pattern under unified leadership, government bodies supervise in accordance with law and regulations, inspection agencies provide technical checks, enterprises bear overall responsibility, and the public take part in supervision.

The fourth stage (2013–present) is the reform period. In June 2013, the Standing Committee of the National People's Congress reviewed and approved the *Special Equipment Safety Law*, completing the five-level legal structure for special equipment safety supervision, namely, law, regulations, rules, technical codes, and standards. The new supervision mechanism was finally established, which was constituted by government supervision bodies, inspection agencies, and social organizations. Special equipment safety supervision function reform, administrative license reform, and inspection reform continue to make progress, and special equipment inspection agencies integration begin to take shape.

(3) The Current Status of Special Equipment Safety Supervision Work

Special equipment safety supervision mechanism. AQSIQ is the competent government body responsible for special equipment safety supervision. AQSIQ is a ministry directly under the State Council in charge of national quality, metrology, special equipment safety, entry-exit commodity inspection, entry-exit health ***quarantine***, entry-exit animal and plant quarantine, import and export food safety, certification, accreditation, and standardization. It also has administrative law enforcement duty. AQSIQ sets up 17 internal departments (bureaus) including Special Equipment Safety Supervision Administration Bureau (SESA). In addition, AQSIQ manages two committees, 17 affiliated entities, and 14 industrial associations. It vertically manages 35 entry-exit inspection and quarantine bureaus throughout the country, and leads the provincial Quality and Technical Supervision Bureaus of the whole country as shown in Figure 14-3.

SESA carries out special equipment safety supervision and administration, and oversees special equipment design, manufacture, installation, alteration, maintenance, use, testing, inspection, import, and export. Within its scope of authority, SESA organizes investigation of special equipment accident and handles statistical analysis. It supervises and manages qualifications of special equipment inspection agencies, inspectors, and operators and monitors the energy saving performance of high energy-consumption special equipment. SESA has seven internal divisions. They are comprehensive division, inspection and information management division, boiler and pressure vessel safety supervision division, pressure pipeline and gas cylinder safety supervision division, elevator and hoisting machinery safety supervision division, special equipment energy saving supervision division, and accident investigation division.

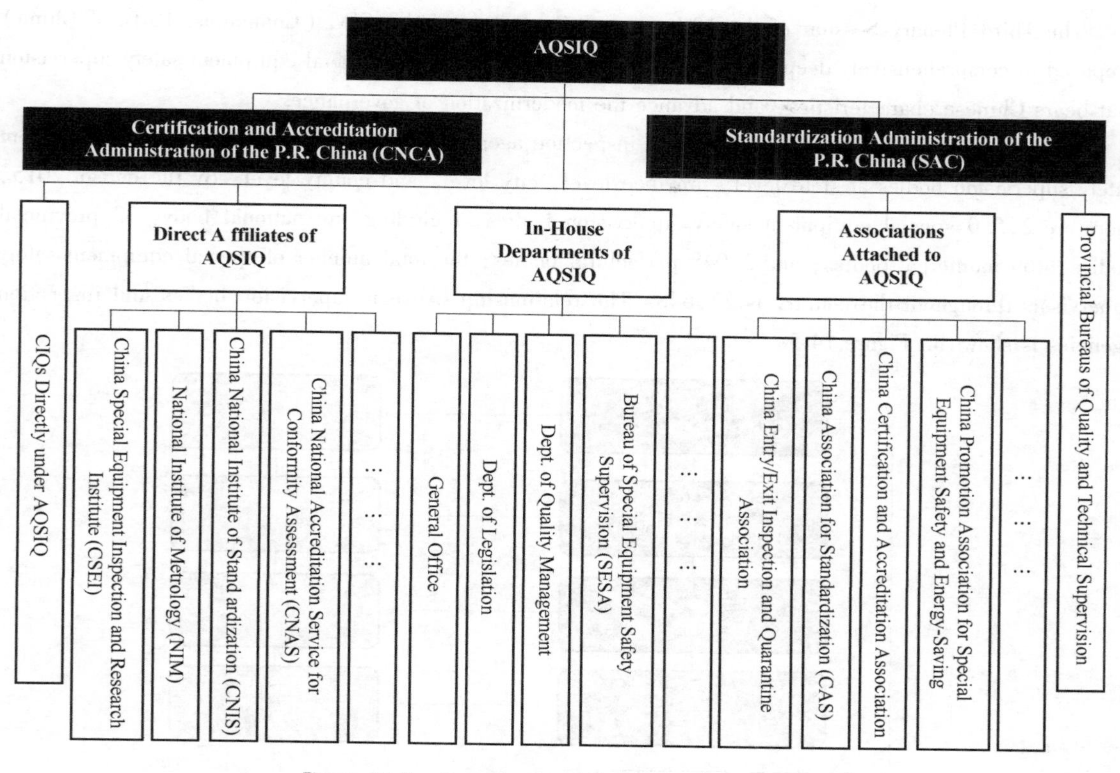

Figure 14-3 Organizational Structure of the AQSIQ

Before the beginning of the reform and opening-up, special equipment safety supervision relied on administrative supervision only. After the reform and opening-up, by referencing the experience of developed countries and based on China's basic national conditions, the "double track" system of special equipment safety supervision gradually comes into being, and safety supervision bodies work side by side with technical inspection agencies and pay equal attention to afterwards treatment and beforehand prevention. With the establishment and perfecting of the market economic system and the deepening of the reform and opening up, special equipment safety supervision changed to "troika" mode, in which supervision bodies, inspection agencies, and social organizations all play a part as shown in Figure 14-4.

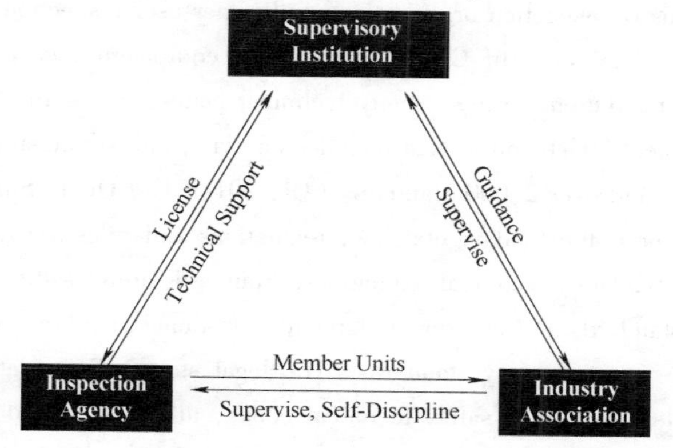

Figure 14-4 China Special Equipment Safety Supervision System

The Third Plenary Session of the 18th Central Committee of the CPC (Communist Party of China) proposed to comprehensively deepen reform, and build a new pattern for special equipment safety supervision that bears Chinese characteristics, and advance the modernization of governance.

Special equipment supervision bodies and inspection agencies. China has established special equipment safety supervision bodies at state level, province level, city level, and county level. By the end of 2015, there are 2,550 special equipment safety supervision bodies, including one national body, 32 provincial bodies, 469 municipal bodies, and 2,048 prefectural bodies; the total number of special equipment safety supervisors throughout the country is 23,648. The relationship between supervision bodies and inspection agencies is shown in Figure 14-5.

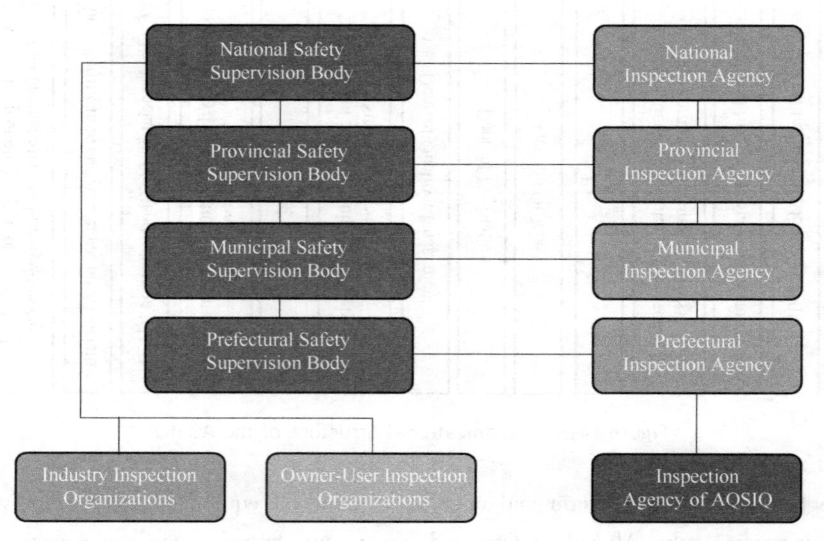

Figure 14-5 Relationship between Special Equipment Safety Supervision Bodies and Inspection Agencies

Inspection belongs to the high technology service industry, and it provides technical support for special equipment safety supervision. Special equipment inspection bodies are mainly composed of two groups: One is the government affiliated inspection agencies; the other includes enterprise owner-user inspection organizations, industry inspection organizations, and other social forces. By the end of 2015, there are a total of 485 comprehensive special equipment inspection bodies, of which 295 are government ***affiliated*** inspection agencies, and 190 are industry inspection organizations and owner-user inspection organizations.

Regulation and standard system. In China, the special equipment legislation and standard system includes five levels: law, regulations, rules, safety technical codes, and standards (Song, Shi, & Xie, 2005). The special equipment legislation system includes one law, one administrative regulation, 10 rules, 121 safety technical codes, and over 2,000 standards (Qi, 2014; Qi, Qian, Shi, & Zhang, 2014). The *Special Equipment Safety Law* is the fundamental law, regulations and rules are refinement of the law, and safety technical codes are mandatory technical regulations. Standards are classified as national standards and local standards, industry standards and enterprise standards. National standards and industry standards are further divided as mandatory or voluntary standards. The legal status of mandatory national and industry standards is almost equivalent to that of safety technical codes, all being issued by competent government departments in the form of notice. Mandatory national and industry standards focus on the technical specification and its implementation method, and safety technical codes focus on the basic safety requirements and management measures. At present, China is promoting the integration of safety technical codes and the development of standardization. Hundreds of safety technical codes are planned to be integrated into dozens of comprehensive safety technical codes, and ***mandatory*** standards of special equipment will be phased out. To be in line with international practice, China encourages the development of social organization standards.

2. Experience Summary and Problem Analysis

(1) Basic Experience

China has applied safety supervision to special equipment for over 60 years, referencing the mature experience of developed countries, and by using the accident causation theory and the system safety theory, China established the basic supervision mechanism which includes administrative licensing and supervisory inspection, and perfected the whole process safety supervision system. The following five basic experiences can be summarized after reviewing the development course of special equipment safety supervision.

The unified leadership of the government provides guarantee. Every year, the State Council issues special equipment accident control targets to provincial governments. Governments at all levels supervise and notify major potential safety hazard. In-service industrial boiler energy efficiency testing, as one of the energy-saving responsibility evaluation element, has been ***incorporated into*** the performance assessment system of the government. Special equipment safety supervision bodies' functions have been unified; the procedures, time limit, and requirements of special equipment administrative licensing and supervisory inspection have been clearly defined (Tao, Shen, Yu, & Li, 2015). The unified leadership of the government has laid a solid foundation for the development of special equipment safety supervision and energy saving.

The overall process supervision and classified supervision continue to improve and play fundamental function. Whole process safety supervision to special equipment is an effective measure to control accidents and reduce safety risk. This measure conforms to China's national conditions and the basic law of safety. On the basis of risk analysis, China applies classified supervision to equipment, entities, and areas. Critical equipment field is the focal point for supervision. Supervision differentiations are applied to different entities according to actual local conditions. Different supervision measures have been deployed for different areas. Thus the allocation of resources is optimized and has received good results.

Supervision and inspection work together to improve effectiveness. Setting up special equipment safety supervision bodies ***parallel with*** inspection agencies at the same jurisdictional level improves the organizational construction. Inspection agencies are technical supporting organizations, and they shall perform inspections in accordance with law. Supervision bodies and inspection agencies work independently while closely with each other, bringing respective advantages into full play to produce synergy effect.

Science and technology research provide support for safety supervision. Special equipment science and technology development is mainly led by government affiliated special equipment inspection agencies, and jointly undertaken by end users, research institutions, institutions of higher learning, and social inspection and testing institutions. During China's 12th Five-Year-Plan period, led by China Special Equipment Inspection and Research Institute, the special equipment inspection industry undertook altogether 245 scientific research projects of various kinds with total investment of 301 million *yuan*, tackled nearly 300 key technologies, and won the national science and technology progress prize twice. Special equipment safety inspection and evaluation innovation team has been awarded as one of the national key innovative teams. China special equipment safety and energy saving technology demand framework for the 12th Five-Year-Plan is as shown in Figure 14-6.

Special Equipment Safety and Energy Saving Technology Demand

Diagnosis and Assessment tech.	Inspection Tech.	Regulation and Standards	Emergency Rescue Tech.	Energy-Saving and Cost Reducing Tech.	Scientific Supervision Tech.
1.Buried pipeline 2.Pressure vesseland pipeline 3.Boiler 4.Amusement device 5.Elevator 6.Lifting appliance 7.Passenger ropeway 8.Non-road vehicle	1.Electromagnetic technology (pulsed eddy current, magnetic flux leakage, magnetic memory, electromagnetic ultrasonic) 2.Sound technology (acoustic emission, guided wave, ultrasonic) 3.Pipeline inner inspection technology 4.Other new technologies (digital radiography, light, etc.)	1.Comparison study of domestic and foreign regulations and standards 2.Material performance test 3.Advanced design method and structure 4.China Special Equipment International Competitiveness Analysis	1.Emergency rescue system and emergency technical platform 2.Pressured sealing technology 3.Emergency protection device 4.Emergency plan system	1.Energy consumption index evaluation method 2.Energy efficiency test method and device 3.The diagnostic methods of energy efficiency 4.Advanced energy saving technology	1.Dynamic monitoring data flow, database and data mining 2.risk based evaluation and supervision system 3.Optimization research and the social impact evaluation based on the requirement of security technology 4.Scientific regulatory model and operation mechanism optimization

Figure 14-6　China Special Equipment Safety and Energy Saving Technology Demand Framework

Build the multi-party cooperation management mechanism to create a good working atmosphere. For special equipment safety supervision, the establishment of multi-party cooperation management mechanism of "unified leadership, enterprises undertake overall responsibility, joint supervision by different government departments, inspections provide technical support, and extensive social participation" plays a positive role in encouraging the whole society to take part in safety supervision.

(2) The Existing Main Problems

With the rapid development of economy, China's special equipment safety work has achieved remarkable achievement, but compared with the developed industrial countries, accident rate is still high, major and extraordinary accidents still occur, and the safety situation is still grim. Under the economic development of the new normal, special equipment safety supervision work still has some deep-seated contradictions and obvious problems. Government supervision effectiveness needs to be improved, market mechanism does not play its role well, the social credit system is not sound, and the social safety awareness is not enough.

New Words and Expressions

supervise [ˈsjuːpəvaɪz]　*v.* 监督，管理，指导

stipulate [ˈstɪpjʊleɪt]　*v.* 规定，保证

promulgate [ˈprɒm（ə）lgeɪt]　*v.* 公布，传播，发表

quarantine [ˈkwɒrəntiːn]　*v.* 检疫，隔离，使隔离

affiliate [əˈfɪlɪeɪt]　*v.* 参加，加入，发生联系

mandatory [ˈmændət（ə）ri]　*adj.* 强制的，托管的，命令的

incorporate into　并入

parallel with　平行，与……比较

Exercises

1. What's the special equipment? Simply explain the special equipment safety.
2. What is special equipment safety technique? What may special equipment safety technique consist of?
3. How to promote the formation of social intermediary organizations of special equipment?
4. What is the main function of Special Equipment Inspection?
5. How to enhance special equipment safety in your opinion?

科技英语翻译技巧——长句的翻译

由于科技英语的特殊性，常常会出现大量的长复合句，用来解释某些复杂的概念。科技英语必须严谨、准确，因此必须带有许多修饰、限定和附加的成分。这些附加的成分会使句子变得复杂并且冗长，甚至有时一个段落就只由一个句子组成。

英语句子的长句动词少，语序灵活；汉语句子短，动词多，语序比较对称。因此，在翻译英语长句时，必须弄清楚原句的结构关系，先了解原文的大意，然后在理解原文的基础上分清句子成分，按照汉语的习惯，将句子的各个分句进行组合。科技英语长句的翻译方法有顺序法、逆序法和分句法。

1. 顺序法

如果长句的语序顺序和汉语的表达方式相近，依次相接，那么就可按照原句的顺序、层次进行顺序翻译。

We have a basic criterion for hotel fire safety—a fire sprinkler system with sprinklers in every room, installed in compliance with nationally recognized standards and then maintained by qualified technicians.

我们对酒店的消防安全有一个基本标准——在每个房间都安装符合国家认可标准的带有喷头的喷水灭火系统，然后由合格的技术人员维护。

Mine fires may from spontaneous combustion, from flame or sparks from various sources, such as defective equipment, shot-firing, smouldering cigarettes and matches, or from heat due to friction as developed by running rope and belts.

矿井火灾也可能由自燃引起，或引燃自各种来源的火焰或火花，如损坏的电力设备、放炮、引燃的香烟、火柴，或者引燃自运行绳和传送带摩擦产生的热。

2. 逆序法

在多数情况下，英语的表达方式和汉语不同，长句的结构顺序和汉语语句的结构顺序相反，这就要进行逆序翻译，从英文原句的后面向前翻译。

Iron rusts at its exposure to the open air on account of the corrosion made by the destructive chemical attack of a metal coming into contact with suck media as air, water and moisture.

金属在接触空气、水和湿气等介质时，会受到破坏性的化学侵蚀，由于这种腐蚀，铁暴露在露天时会生锈。

Scientists are learning a great deal about how the large plates in the earth's crust move, the stressed between plates, how earthquakes work, and the general probability that given place will have an earthquake, although they still cannot predict earthquakes.

尽管科学家仍不能预测地震，但对地壳中的大板块如何运动、板块间的压力大小如何、地震是如何发生的、特定地区发生地震的一般概率为多少，他们了解得越来越多。

3. 分句法

英语的长句常常由各种形式的分句或短语构成，当这些分句或短语与所修饰词的关系不是很明确、各自成句时，可根据汉语的习惯，将一个分句或短语独自翻译，再根据原文的逻辑关系进行排序，使用连词连接起来。

The amendments include revisions to Section 3 Methodology, including the addition of a paragraph outlining the need for data on incident reports, near misses and operation failures to be reviewed objectively and their reliability, uncertainty and validity to be assessed and reported.

这些修订包括修订第 3 条方法论，增加了一段，概述了事故报告数据的必要性，侥幸时间和操作失败应客观审查，对其可靠性、不确定性和有效性进行评估和报告。

Falsework should be constructed, or adapted, so as to be suitable for the purpose for which it is used. It should be strong enough and stable in use; damaged components should not be used; and different proprietary components should not be mixed, unless expressly approved by the designer.

为了适应脚手架的实际用途，需要制作或改装脚手架。脚手架在使用中应确保足够牢固和稳定；损坏的部件不能使用；不同的专有部件不能混用，除非另有设计者明文批准。

New Development of Special Equipment

Special Equipment Safety Technology has been developed nearly 30 years. It has accumulated a number of achievements, achieved a kind of career, created a generation of elite, and led the sector progress. The technology played the role of technical support for our pressure special equipment accident rate in 20 years (1987–2007) that declined rapidly 18 times. It has made great social and economic benefits.

1. Innovation in Risk Assessment of Special Equipment

Special equipments mean equipments that are dangerous and relating to lives such as boiler, pressure container, elevator, crane, and the passenger ropeways. There is already a method of weighted average analysis used in the ***Artificial Neural Network*** (***ANN***) model of the risk assessment for special equipments, whose time is too long as the adopted weights are all based on expert experience. So a risk assessment model based on Fuzzy Analytic ***Hierarchy*** Process (F-AHP) and ANN is proposed. Firstly, to carry on the level analysis to all possible influences factor using the F-AHP method; calculates the weight of each influence factor which occupies in the influence special equipment safe state; then according to the weight choose the most influential factor as the input of ANN model. Finally use the risk assessment model on the elevator, the assessment time used is less than former ANN model, and the accuracy is not less than former ANN model.

Another method is based on fuzzy neural network. The evaluation indexes of special equipment are proposed according to its features and they include all chains of life of ***amusement*** ride. According to fuzzy safety evaluation, the evaluation factors and weights are described and the evaluation model is set up. The fuzzy neural network model of safety evaluation of special equipment is built according to Back Propagation (BP) neural network. The weights of evaluation factor are ***optimized*** based on study and training. The fuzzy neural network system is developed and with the example of giant wheel, the weights of evaluation factors are optimized.

2. Innovation in Non-Destructive Testing

Special equipments are very important equipments with potential danger, which are widely used in industrial production and people's daily life. ***Non-destructive Testing*** (***NDT***) plays one of the most important roles in insuring manufacture quality and operation safety of special equipments. The NDT techniques are used for raw material, manufacture and in-service testing of special equipments, which includes radio graphic testing, ***ultrasonic*** testing, magnetic particle testing, dye penetrant testing, eddy current testing, acoustic emission testing, metal magnetic memory testing and magnetic flux leakage testing.

Recent inspections of pressure ***vessels*** have shown that there are a considerable number of cracked and damaged vessels in workplaces. Cracked and damaged vessels can result in leakage or rupture failures. Rupture failures can be much more catastrophic and can cause considerable damage Acoustic Emission (AE) measurements have become a reliable and standard method of NDT for pressure vessels. AE is a meticulous Non-Destructive Examination (NDE) method that exposes ***deficient*** areas in pressure vessel integrity. It is the only NDE method able to assess volumetric integrity during a vessel pressure test period. When AE is used as a primary examination method during hydrostatic testing, it supports all other NDE methods to life and property.

The joint use of fractography, potential difference as well as acoustic emission confirms the identification of the acoustic signatures of the various phenomena of damage in the base metal, the HAZ and the weld of CT specimens. The count, the amplitude, the rise time, the average frequency and the duration are the important parameters for AE identification of the different crack ***propagation*** mechanism. We can also identify the zone of defect propagation, but the AE parameter used for that depends on the kind of steel.

3. Innovation in Safety Monitoring

The technology of virtual instrument is applied in working condition inspection, and fault diagnosis of well winch, crane, elevator and the other special equipment. By LabView graphic program software, the signal could be calculated, analyzed and processed so as to realize different tests. This technology is a combination of electronic, computer, monitoring and mechanical engineering technologies. It is a technological renovation in safety monitoring of modern equipment. According to the ***appraisal*** by Henan Provincial Economic and Trade Commission, it successfully fills in the blank in safety monitoring for domestic special equipment. It has social and economic significance in ensuring occupational safety and improving the reliability of equipment operation.

The elevator safety controls system based on PLC is developed, and the system is in a stable and reliable state through the actual test elevator, as one of the main means of transportation in buildings, provides people with safe, fast and comfortable vertical transportation service. Owing to the large population in China, the influence of elevator to people is very significant being accompanied by the appearance of intelligent buildings, the buildings tend to become higher and larger, which made tractive elevator control system become increasingly complex. On the one hand, the widespread usage of the PLC and VVVF made the elevator more comfortable, safe and intelligent. On the other hand, they require the control system to be more sophisticated. Delivering PLC to the core of the elevator control system could meet the demand that the skyscraper should ensure a comprehensive safety for elevators. Furthermore, this system has high stability and convenience for maintenance. In addition, the research of the elevator-group-control-system ***algorithm*** and monitoring system are the major coverage in the search of the elevator control system. Because of its good security measures, perfect monitoring function and friendly human-machine interface, configuration software was used extensively in modern production for monitor and control of industrial production process. What's more, we can collect the elevator's data of its status through the software and monitoring or controlling them after analyzing and processing its data to obtain a direct, clear and accurate elevator status and ***alleviate*** the operators' work pressure.

An FBG (Fiber Bragg Grating) sensor is applied in monitor the condition of pressure vessel. In recent years, advanced composite structures are used extensively in many industries such as aerospace, aircraft, automobile, pipeline and civil engineering. Reliability and safety are crucial requirements posed by them to the advanced composite structures because of their harsh working conditions. Therefore, as a very important measure, Structural Health Monitoring (SHM) in-service is definitely demanded for ensuring their safe working in-situ. FBG sensors are surface-mounted on the hoop and in the axial directions of an FRP pressure vessel to monitor the strain status during its pressurization. The FBG sensors could be used to monitor the strain development and determine the ultimate failure strain of the composite pressure vessel.

4. Innovation in Design and Manufacture

In order to avoid the repetition of the accidents and improve the safety, reliability and economy of pressure equipment a platform is suggested for design, manufacture and maintenance of pressure equipment in China based on accidents survey. In other words, some effective precautionary measures are taken at the design and manufacture stage and the design methodology has to be based on service life requirement and desirable risk level. At the service stage some reasonable inspection/monitoring app roaches should be utilized to control risks and ensure the equipment operating safely until its desired lifespan. The concept of risk and life based design manufacture and maintenance proposed herein has important significance for improving and perfecting the codes and standards for design, manufacture and maintenance of Chinese pressure bearing equipment, enhancing the life and reliability of Chinese pressure bearing equipment and promoting the development of in service maintenance technology that combines safety and economy. Recently in order to save resources, many developed countries in America, Europe and Australia have successively increased permissible stress by adjusting safety factor and strain intensification of austenitic stainless steel and adopted the method of high-strength level steel grades to reduce steel consumption of pressure-bearing equipment, and our country has also proposed similar requirements in its safety technical code which has been promulgated recently and will be put in place. Under the great trend of global weight lightening of heavy-duty press-bearing equipment and seeking for harmony of ***intrinsic*** safety and energy and material conservation, the researches on the design, manufacture and maintenance based on risk and life will be of important significance for ensuring the safety of weight lightened pressure-bearing equipment.

New Words and Expressions

Artificial Neural Network (ANN)　　［电子］人工神经网络
hierarchy [ˈhaɪərɑːki]　　*n.* 层次体系，等级制度，统治集团
amusement [əˈmjuzmənt]　　*n.* 娱乐，娱乐活动，可笑，愉悦
optimize [ˈɒptɪmaɪz]　　*v.* 持乐观态度，乐观地对待，使尽可能完善，使最优化
Nondestructive Testing (NDT)　　无损检测，［试验］非破坏性试验，原位测
ultrasonic [ˌʌltrəˈsɒnɪk]　　*adj.* 超声的
vessel [ˈves(ə)l]　　*n.* 容器，船，管，导管
deficient [dɪˈfɪʃ(ə)nt]　　*adj.* 缺乏的，缺少的，不足的，有缺点的
propagation [ˌprɒpəˈgeɪʃən]　　*n.* 传播，扩展，宣传，培养
appraisal [əˈpreɪz(ə)l]　　*n.* 评价，鉴定，估价，估计
algorithm [ˈælgərɪð(ə)m]　　*n.* 算法，计算程序
alleviate [əˈliːvieɪt]　　*v.* 减轻，缓解，缓和
intrinsic [ɪnˈtrɪnsɪk]　　*adj.* 固有的，内在的，本身的

Unit Fifteen

15

Integrated Coal Mine Safety Monitoring System

Mine disasters which cause great loss to life and property of the people are frequently seen in many countries, especially in the medium-small sized local mines, so the safety production of coal mine has been a weighty topic that affects production of the mine industry, even social stability. Generally, safety accidents mostly occur in the production process of small local mines. Hence, an integrated ***coal mine safety*** monitoring system based on combination of wireless communication and GPRS network is proposed, which bears such features as low price, reliable performance and easy installation & maintenance etc.

The integrated system is composed of various sensors, wireless Radio Frequency (RF) module substations, main control module of GPRS module and the host computer in the management center. In each substation, various underground sensors for gas, wind speed, temperature etc. carry out real-time monitoring. When the data exceed the limited values, an alarm will occur. Compared with data transmission through conventional bus, application of wireless communication and GPRS network technologies reduces investment on underground layout, ensures timely, accurate and rapid transmission of data in all key underground mining zones and improves efficiency of the system.

Main monitoring parameters of coal mine safety are gas information, temperature information and wind information etc. The system, which carries out automatic and continuous monitoring on the relevant underground parameters, continuously collects data of all parameters that require monitoring by various sensors, sends them to the signal filter circuit where the collected data are instantly filtered through ***fluctuation***, and after such circuit treatment as A/D amplification, they are transmitted to the MSP430 type single-chip computer in each substation that made by TI company where qualified data are stored in corresponding memory cells. Signals in the mine are transmitted through wireless data. Due to the fact that there are several working faces in one mine and they are not far from each other, it is therefore designed that all data from various working faces are collected to the substations of the mine through RF technology where they are transmitted to the main control module, and from there the date are submitted to the management center and mobiles of the responsible personnel through GPRS and GSM. In this way, real-time monitoring of all mine substations is obtained through convenient, effective and economical ***transmission*** (see Figure 15-1).

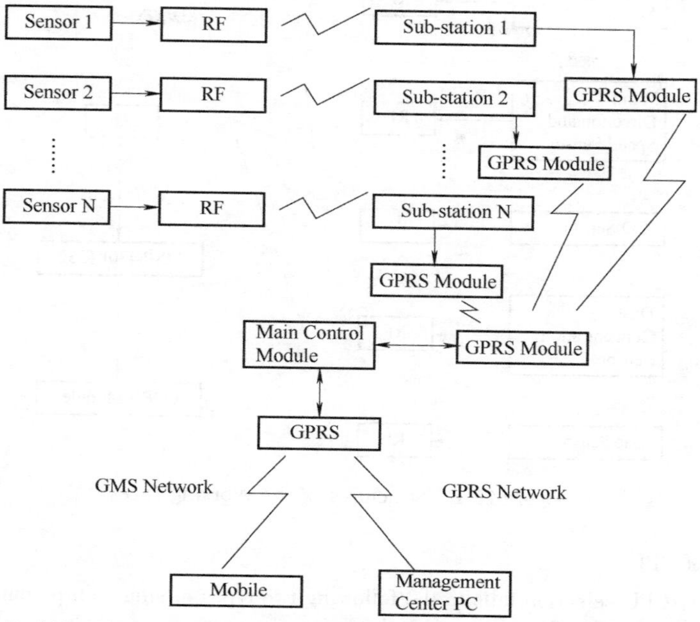

Sensor 1 → RF ~ Sub-station 1 → GPRS Module
Sensor 2 → RF ~ Sub-station 2 → GPRS Module
Sensor N → RF ~ Sub-station N → GPRS Module
Main Control Module ↔ GPRS Module
GPRS
GMS Network
GPRS Network
Mobile
Management Center PC

Figure 15-1 Real-Time Monitoring Through Convenient, Effective and Economical Transmission

1. System Structure

Figure 15-1 shows the general structure of the system. It is seen that the substations carry out real-time ***detection*** of relevant parameters through various sensors installed underground for CH_4, CO, H_2S, O_2, SO_2, NO_2, NH_3, wind speed, negative pressure, temperature etc. Through nRF905 wireless RF module with 433MHz frequency channel, data are transmitted to the substations, then after treatment by single-chip computer, data are transmitted to the main control module near the mine through the GPRS module. The single-chip computer (MSP430F149) in the main control module receives various index data from various substations and transmits real-time data to the host computer of mine and mobiles of the responsible personnel through the GPRS module and GSM network. The GPRS module in the management center is connected to the host computer through serial interface. If the data are normal, it can be set that the data are sent once every other hour while in case of abnormal data (that is, danger might occur), the alarm device is activated and the data are transmitted at once.

Corresponding data collected by various sensors in each working face are sent to the substations through the wireless RF transmission module (nR905), then treatment by the single-chip computer (MSP 430F1232). For structures of the working faces, refer to Figure 15-2.

2. Selection of Hardware

The ***hardware*** system consists of main chips like MSP430 single-chip computer, A/D converter etc., circuit and sensors. As there is a 256B data memory in the MSP430 single-chip computer, no external memory is arranged.

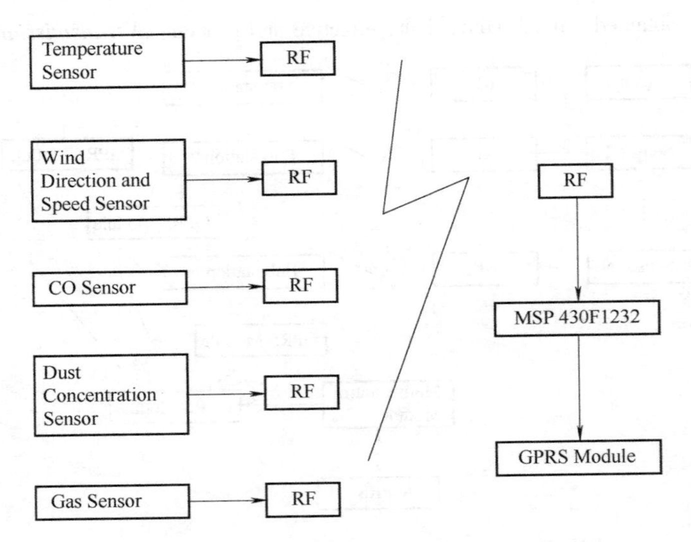

Figure 15-2 Structures of the Working Faces

(1) Selection of CPU

In system design, CPU selection falls to the following two types of single-chip computers, MSP430F149 and MSP430F1232 which are Mixed Signal Processors with low energy-consumption and high performance made by TI Company. Power supply is in low ***voltage*** from 1.8–3.6V. The CPU can be set to power-saving mode with programs being activated through interruption method. Operating ***ambient*** temperature is between -40℃ and +85℃. MSP430 is a 16-bit processor with RISC structure, there are 16-bit registers, 10 kb RAM, 48 kb + 256 kb flash chip, with two 16-bit timer, a 12-bit A/D converter and two 12-bit D/A converter, two serial communication receive and transmit port. The most prominent feature is the low power consumption, can work in the 1.8 V to 3.6 V range. In the "Active" state, the operating current is only 330μA, in the "Standby" state, the operating current is only 1.1μA, in the "Off" state, the operating current is only 0.1μA, it also offers five kinds of power saving mode can be selected by the software. As the coal mine environment, bad, serious interference, so the hardware and software anti-interference are concluded in the design phase. Such as the design of the software trap instruction ***redundancy***, software "watchdog" and so on. In order to ensure stability, designed a special watchdog circuit will automatically reset can also be manually reset.

(2) Selection of Analog-Digital Converter

The 16-bit A/D conversion chip, AD 7718 made by AD Company with built-in digital filter that possesses interference suppression is selected. The analog input of AD7718 can be connected to sensors directly and its digital interface (standard SPI interface) is connected to single-chip computer directly.

(3) Selection of Sensors

Sensors are the primary link in automatic detection and automatic control, so it is necessary to select proper sensors. In this system, different sensors are used to collect data from different objects.

Most of the domestic coal mines are gas mines. That's why casualties caused by gas explosion accidents occupy over 50% of all casualties in major accidents. For detection of gas, low power gas sensor UL-264, ***a kind of*** heat conduction type gas sensor, is selected of which the working principle is that when the detected gas doesn't change in composition but in density, the density is in linear proportional relationship with the change of resistance. Within the full-scale range (0–100%), there is a good linearity and a very limited non-linearity error.

TGS2442 is very sensitive to CO, so if CO exists, the electrical conductivity of the sensor will increase with the density increase of the gas in the air; it's not easily affected by temperature.

Dust is detected by dust concentration sensor. Measurement of dust concentration is made through the principle of light scattering. It is mainly made up of ***optical*** measurement chamber, extraction pump, flowmeter, filter and data processing and control circuits mainly composed of PIC16F887 single-chip computer.

Temperature is detected by DS18B20 digital type temperature sensor, a new type made by the US DALLAS Company with single bus technology. Both input and output of the sensor are digital signals and are connected to the external through serial interface mode.

Wind speed is detected by the three-cup sensor, of which the wind-cup that drives the rotation of the photoelectrical encoder is made of carbon fiber reinforced plastic. The pulse signals are produced by a photoelectrical conversion device of which the conversion circuit is made of porous metal and photoelectrical tube that produces frequency signal in direct proportion to the wind speed.

3. Design of RF Module

The RF technology is a type of wireless communication technology that transmits information by taking electromagnetic waves as the carrier waves. Currently, most RF chips work in the 433/868/915/2400MHz frequency ranges. In the high frequency channel of 2.4GHz, the communication rate between the server and the terminals can reach 12Mbps; besides this frequency channel is under no wireless control in most countries. The transmission chips of different substations can be distinguished by the address codes assigned to the RF, thus realizing simultaneous data transmission from multiple transmission modules to the main module.

nRF905, a relative more excellent type of RF chip compared with its equivalents, is integrated from such modules as power management, crystal oscillator, low-noise amplifier, frequency ***synthesizer*** and power amplifier etc. with the Manchester encoding/decoding being accomplished by the built-in hardware of the chip. FSK (Frequency-Shift Keying) modulation mode is adopted for direct reception and transmission. Through the serial interface, it is connected to the micro-controller and works in 433/868/915MHz with the function of carrier waves detection. The power consumption is low and through the POWERDOWN mode, power saving is easily obtained.

The ***antenna*** which is the interface that the electromagnetic waves spread along the transmission route and in the space is both a passive device and reciprocal device. That is to say, the same design can be used as transmitting antenna as well as receiving antenna. For a good antenna, such factors as frequency band, impedance and gain shall be taken into consideration. In application, part of the circuit antenna can be of the loop type in differential connection, that is to distribute the antenna on the PCB board, in which way, the system volume can be reduced. In this mode, the grounding network layer is at the bottom and the circumstance of the components on the board top have also grounding surface to ensure sufficient grounding of adjacent components.

4. GPRS Module Design

GPRS is one of realizations of the GSM Phase2. 1 (Global System for Mobile Communications) Specification. The GPRS is of the same frequency band, bandwidth, burst structure, wireless modulation standard, hopping rules and TDMA (Time Division Multiple Access) frame structure, so the current BSS (Base Station Sub-System) provides overall GPRS coverage from the very beginning. There is no need for additional giant data terminals for SMS and information communication can be realized with mobiles. Siemens MC series, Motorola G-series, Sony Ericsson's GR series GPRS wireless module is common in the market. We selected the SIM300C module with built-in TCP/IP protocol stack made by SimCom Company, in which standard AT orders and rich instruction sets are provided to the customers who can integrate them into various data terminals in an easy way. Its advantages lie in that it always maintains rapid and stable wire connection and data transmission with fast GPRS technology. According to actual tests, the dropdown probability is very low. Control mode for input/output of SIM300C can be in such way that makes the external control application provided by external input/output interface more efficient. The embedded control unit is accessible to application fields like alarm, instrument collection and application release and supports language programming. It has its own TCP/IP protocol stack function, can open its internal CPU resource and owns all necessary functions for M2M communication (Figure 15-3).

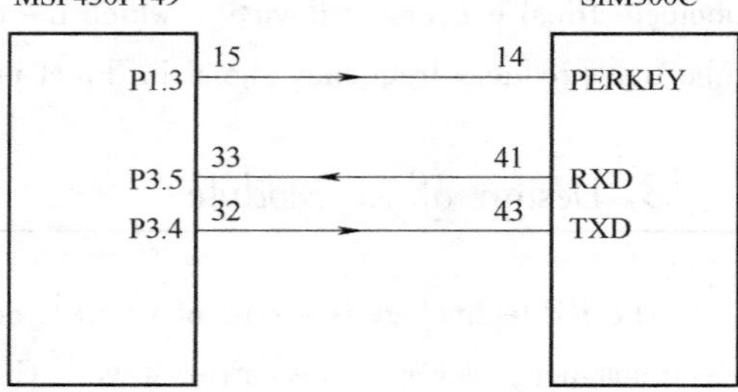

Figure 15-3 GPRS Module Design

For the rate matching, at present, China Mobile GPRS mainly use the CS-2 channel coding scheme, providing 1–4 GPRS channel (PDCH, Packet Date Channel) in the up and down channel each. At first, the system need at least one static PDCH, and then adjust the additional number of PDCH base on the flow. According to the principle of voice first, the additional PDCH will be allocated voice service, to ensure the quality of GSM network. Therefore, the actual bandwidth of GPRS is 13. 4Kbps–53. 6Kbps. In the GPRS network, using the method of distribution of asymmetric uplink and downlink channels, uplink use the low bandwidth and downlink use the wide bandwidth, usually 1 +2, 1 +3, 2 +4 and so on, that was mainly set up for customers to access the Internet. But in industrial applications, the situation is changed; the bandwidth need of uplink is more than the downlink. So our system use the special GPRS module with the Class 10 support, can provide 4 downlink channel and 2 uplink channel. In this way, the uplink and downlink bandwidth can be 120kb/s and 50kb/s.

After extensive researches on relevant theories and technologies on various early warning systems for mine gas disasters, this relatively new, advances and practical detection system is proposed, which will exploit its advantage in full in practical application, leading to the improvement in decrease of mine disasters.

New Words and Expressions

coal mine safety 煤矿安全，煤炭安全，矿井安全

fluctuation [ˌflʌktʃʊ'eɪʃ (ə) n] *n.* 波动，脉动，踌躇，彷徨变异

transmission [trænz'mɪʃ (ə) n] *n.* 传播，传递，变速器，传送
detection [dɪ'tekʃ (ə) n] *n.* 探测，发现，侦查，察觉
hardware ['hɑːdweə] *n.* 硬件，硬件设备，五金制品，机器
voltage ['vəʊltɪdʒ] *n.* 电压，伏特数
ambient ['æmbɪənt] *adj.* 周围环境的，周围的，产生轻松氛围的
redundancy [rɪ'dʌnd (ə) nsi] *n.* 多余，解雇，累赘
a kind of 一种，一类，某种，有一点
optical ['ɒptɪk (ə) l] *adj.* 视力的，视觉的，光学的，有助于视力的
synthesizer ['sɪnθəsaɪzə] *n.* [电子] 合成器
antenna [æn'tenə] *n.* 触角，触须，〈比喻〉直觉

Exercises

1. What is the characteristic of integrated coal mine safety monitoring system?
2. What is System Structure? What is the basic operation principle of System Structure?
3. Explain the RF (Radio Frequency) technology simply?
4. What does hardware consist of? How to detect dust, temperature and wind speed?
5. What is the characteristic of GPRS Module?

科技英语翻译技巧——it 的用法

在科技英语中，含有 it 的句型极为常见，可构成多种不同的结构。在英语中，it 可用作人称代词、指示代词、无人称代词、强调代词和先行词。

1. 人称代词

当 it 用作人称代词时，用来代替上文提到过的无生命的事物或事件，可翻译为“它”“其”，也可重复前面出现过的名词，也可根据情况省略不译。

Potential energy, though it is not so obvious as kinetic energy, exists in many things.
势能虽然不像动能那样明显，但它存在于许多物体之中。(译成“它”)

When it is not use, cover the device with a piece of cloth in order to keep the dust out.
机器不用时，请用布盖好，以免落入灰尘。(译成“机器”)

The aeroplane is supported by the wings; it is propelled by the power plant; it is guided by its control surfaces.
飞机靠机翼支撑，靠动力装置推进，靠操纵面导向。(省略不译)

2. 指示代词

当 it 做指示代词时，相当于 this 和 that 在句中的作用，用来代替前面出现过的名词或句子，可翻译成“这”，指代出前面的内容，或者省略不译。

Motors must be oiled once or twice a month. It depends upon how much they are used.
电动机每月必须加一次或两次油，加油次数取决于电动机使用的频度。(译成“加油次数”)

It is because a conductor carrying a current is surrounded by a magnetic field.
这是因为载流导体周围有一个磁场。(译成“这”)

The birth of the baby from a test-tube was a very difficult process, but it was successful.

婴儿从试管中诞生是个极其困难的过程，但毕竟还是成功了。（省略不译）

3. 无人称代词

当 it 用作无人称代词时，用来表示一般的现象、规律、时间和情况，此时 it 没有词汇上的意义，一般可省略不译，也可以选择适当的名词作为主语。

It is all pitch dark in the depth of the sea, yet life abounds there just as well.

海洋深处一片漆黑，可是照样有大量生命存在。（省略不译）

It is getting very hot.

天气燥热起来了。（译成“天气”）

4. 先行词

it 用作先行词时，可充当主语形式或者宾语形式。

（1）形式主语

it 放在主语的位置，作为动词不定式短语、动名词短语或从句的形式主语，省略不译。

1）代替动词不定式。

With the wide variability of values noted, it is understandably difficult to derive a “typical” amount, but a typical amount might be estimated at about $2500 per year.

由于奖金额相差巨大，所以很难推导出一个普遍的奖金金额，这是可以理解的。但是，据估计，每年 2500 美元是一个有代表性的数额。

2）代替主语从句。

As the cost of navigation systems continues to come down, it is expected that these systems will start to appear in developing countries as well.

由于导航系统的费用持续下降，预计这些系统也将开始应用于发展中国家。

3）代替动名词。

It saves an enormous amount of time figuring on an electronic calculator.

用电子计算器进行运算可以节省大量的时间。

（2）形式宾语

it 放在宾语的位置上，作为动词不定式短语、动名词短语或从句的形式宾语，省略不译。

1）代替动词不定式。

Scientists have proved it to be true that the heat we get from coal and oil comes originally from the sun.

科学家已证实，我们从煤和石油中得到的热都来源于太阳。

2）代替动名词。

They think it necessary solving the problems one by one.

他们认为需要逐一解决这些问题。

3）代替从句。

The invention of radio has made it possible for mankind to communicate with each other over a long distance.

无线电的发明使人类有可能进行远距离通信联络。

5. 强调

it 在强调句中，意为“正是，就是，只是”。

About how many elements is it that make up most of the substances we meet in everyday life?

到底大约有多少种元素构成了我们生活中所见到的大多数物质？

It was not until 19th century that heat was considered a form of energy.

直到19世纪，热才被认为是能的一种形式。

It is a company's culture that makes it safe (or not safe) for a person, division or the whole company to raise issues and solve problems, to act on new opportunities, or to move in new, creative directions.

企业的文化在企业提出和解决问题的过程中，在抓住新机遇实施行动的过程中，或者在开拓新的、创造性的方向的过程中，会使组织的个人、部门或整个企业处于安全状态或不安全状态。

Reading Material

Development of Underground Mine Monitoring and Communication System Integrated ZigBee and GIS

Underground mine safety and health remain challenging issues in the mining industry. Death toll statistics in China's coal mines have gradually reduced these years but fatality still occurs. The number of occupational mining fatalities in the United States' underground metal mines has fluctuated from 40 to 46 during the years 2001–2010. Most importantly, 33.8% of the deaths have resulted from ignitions and explosions of gas or dust in underground mining. In April 2014, two men were killed when a wall collapsed in an underground coal mine in New South Wales, Australia. Human errors were concluded from reports as the most significant reasons for mining fatalities. Thus, safety is always a significant concern in mining operation. Some studies have recently focused on improving the health for underground miners. Laney and Attfield have drawn attention to the fact that the ***prevalence*** of coal workers' pneumoconiosis or progressive massive fibrosis increased from 1990 to 2000 among the United States underground miners. Therefore, specific consideration of both safety and health issues deserves priority in mine operation management and engineering designs to provide and maintain a safe and healthy workplace. In response to these challenges, mine automation by new technologies such as Wireless Sensor Network (WSN) assisted with Geographic Information System (GIS) has been widely utilised in underground mines to enhance safety and health, productivity and reduce operational costs.

The underground WSNs consist of a few to several hundred nodes between a surface gateway and specified sensor nodes at underground levels. ZigBee based on IEEE 802.15.4 protocol is a new wireless sensor technology which has more benefits than other WSNs for underground monitoring and communication systems. Even though ZigBee technology provides only a low data rate, its benefits are low power consumption, very cost-effective nodes, network installation and maintenance. It is also capable of providing networking applications for data transmission between nodes (node to node relays) with high performance based on many wireless hops. It does not require any access point or central node to transmit data between clusters. Significance of ZigBee in underground mines compared to other WSNs was evaluated in the recent publication of authors.

GIS is a new technology used for spatial data analysis in order to capture, store, analysis, manage, and present data that is linked to locations. GIS allows users to view, understand, question, interpret, and visualize data in many ways that reveal relationships, patterns, and trends in the form of maps, globes, reports, and charts. Web-GIS is an inevitable trend which helps solve the problems of spatial information integration and sharing in technical aspect of web media. Recently, researchers have technically focused on the GIS supported for the management of emergency and unsafe conditions.

In this study underground safety and health concerns are significantly mitigated based on the system integration which incorporates ventilation management and emergency message texting. The system integration based on the development of ZigBee nodes is introduced to sense the underground mine environment, to regulate ***ventilation system*** and to communicate between surface offices and miners. Therefore, reduced power consumption, near real-time monitoring of the environment and ***bilateral*** communicating between surface and underground personnel are achieved. Experimental tests were carried out to verify network reliability and security of the packet delivery in underground mines. The architecture of underground monitoring and communication for the system integration is illustrated in Figure 15-4. Temporal ZigBee data including messages and environmental attribute readings such as temperature, humidity and gases concentration are transferred to GIS management server in the surface control centre. The transmitted data are received and stored by ZigBee program then provided for manipulation in the control centre. Risk situations are immediately identified and responded through a logical process of data analysis in the GIS management server before reaching dangerous (unsafe) levels and accidents occurring. The ventilation system management is also used for the workplace health and safety compliance and the optimisation of mine site power usage.

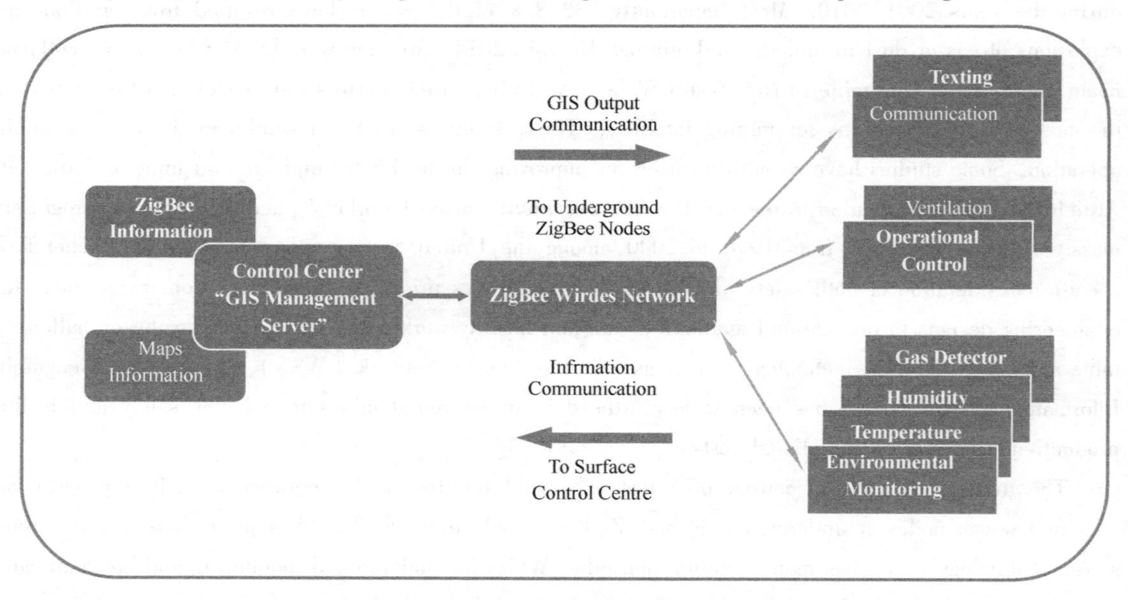

Figure 15-4 Architecture of Monitoring Is Organized and Communication System in Underground Mines

The remainder of this study is organised as follows. The fundamental knowledge of ZigBee technology and GIS are first described. Then, the implementation and structure of system integration are demonstrated. Finally, the strategic process of combining ZigBee data and map information through the GIS management server are modelled for monitoring, communication and controlling the environmental attributes in an underground mine.

1. System Structure

(1) Wireless Network Setup

The entire system of the tested underground WSN is composed of different ZigBee nodes such as coordinator, routers and end devices. These products were developed in collaboration with Tokyo Cosmos Electric Co., Ltd. The JN5148-EK010 kit (Jennic) stacks were employed to create ZigBee network. The wireless network initially is created by coordinator (gateway) to join other nodes. A ZigBee coordinator connected to laptop (PC) using in the experiments is illustrated. Bilateral communication was provided between the coordinator and end devices to send and receive messages and readings instantaneously taken by their sensors. Routers with the ability of sensing the environment were employed to relay communication through the network. In addition, sending and receiving messages and remote control of ventilation fans are enabled by the surface coordinator based on the designed software.

To setup WSNs, power consumption and high reliability of packet delivery are the most concerns. For the former case, ZigBee nodes are ***configured*** to transmit data in longer periods when the mine is in safe and transient conditions which is caused to extend the life of batteries. In latter case, different time intervals are considered for data delivery of environment sensing to avoid network congestion and possibility of packets loss. The power usage of direct and alternating currents (DC/AC) for the ZigBee nodes (except the coordinator) were designed to operate under battery and mine site power supply, respectively. Thus, alternating currents power usage is resulted in the extension of battery life, and ZigBee nodes are enabled to continue long-time data telemetry during power outages at any accident. The ZigBee nodes can last a few days to several months, depending on their data rate and applications.

(2) Sensing Environment

The safety and health of coal and metal/non-metal mining operations were raised considerably as the result of the wireless monitoring of environment attributes. Digital temperature-humidity compound sensor on-board of each JN5148 with advanced sensitivity and long-term stability for mine sites is utilised in the system. Methane, oxygen, CO_2, CO, NO_x and SO_2 concentration sensors (readers) are easily connected to ZigBee nodes to sense the environment. The sensors were configured the single-line communication to transmit real-time data to the nodes. The measurement of CO_2 concentration in this study was considered to manage safety and health risks nearby coal strata in coal mines or fumes-filled spaces in metal/non-metal mines.

(3) Text Messaging Operators

Developed ZigBee nodes are enabled to connect with laptops and mobile phones for sending and receiving text messages. Figure 15-5 illustrates portable radio stations to connect laptop (tablet) which are designed to be placed in an underground refuge chamber, and mobile phones for emergency purposes. The radio station is getting a significant role for wireless communication with surface operator during accidents particularly when cable damage or power outage occurs. Even though, its primary role is the remote control of ventilation fans. ZigBee nodes were placed in the boxes to minimise environmental effects on their operation.

(4) Ventilation Control

Air ventilation deficiency in underground mines is a critical issue to the occupational safety and health of mine personnel. Moreover, optimisation of the fans power consumption to supply underground fresh air is

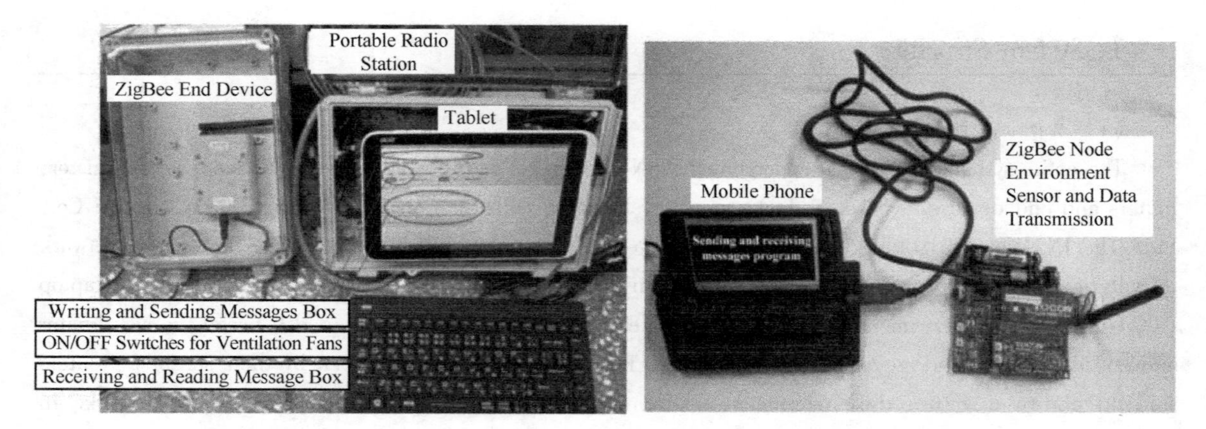

Figure 15-5 Portable ZigBee Radio Stations Providing Communication with Surface Laptop and Miners' Mobile Phones

considered in ventilation system design. Therefore, adding ***auxiliary*** fans to the ventilation system is economically required to improve air quality during hot seasons, blasting, any gas leakages and increase of exhaust fumes. In the proposed system, remote and automatic controls of auxiliary fans were programmed with the software installed on PCs (laptop and tablet) located at the surface office and refuge chamber. Special computer interface with ON/OFF switches and receiving/sending messages installed on PCs is illustrated in Figure 15-6. In the computer interface, separate command icons were designed for each auxiliary fan which gives the user the ability of the ventilation system control.

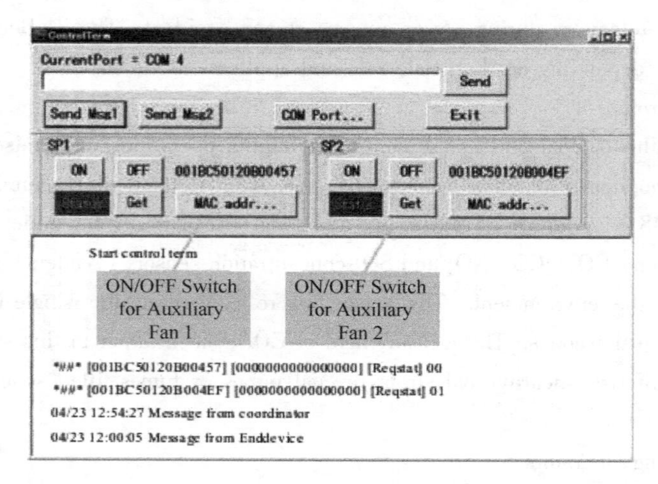

Figure 15-6 Designed Computer Interface to Switch ON/OFF the (Auxiliary) Fans and Receiving/Sending Messages

2. Data Management Server Using GIS

The prototype model developed in this study relies on ZigBee data and geo-processing data of GIS. Data management server was developed on ESRI's established ArcMap 3D software, part of the ArcGIS software package.

(1) Input Data

The first step of our designed management server is to communicate with the outside world to receive required information. Figure 15-7 illustrates data flow sheet and the variety of input data for the GIS management server. Input datasets in the database are comprised of map information, ZigBee nodes data, ZigBee text messages, ZigBee node positions, ***threshold*** limit values and contact details. Map query is primary process of maps to merge and display required features in GIS server to represent the fundamental layers of underground tunnels, geographically. These layers are revised according to the progress of underground mining activities. Then, other input data is analysed and located on the layers for further ***manipulation***.

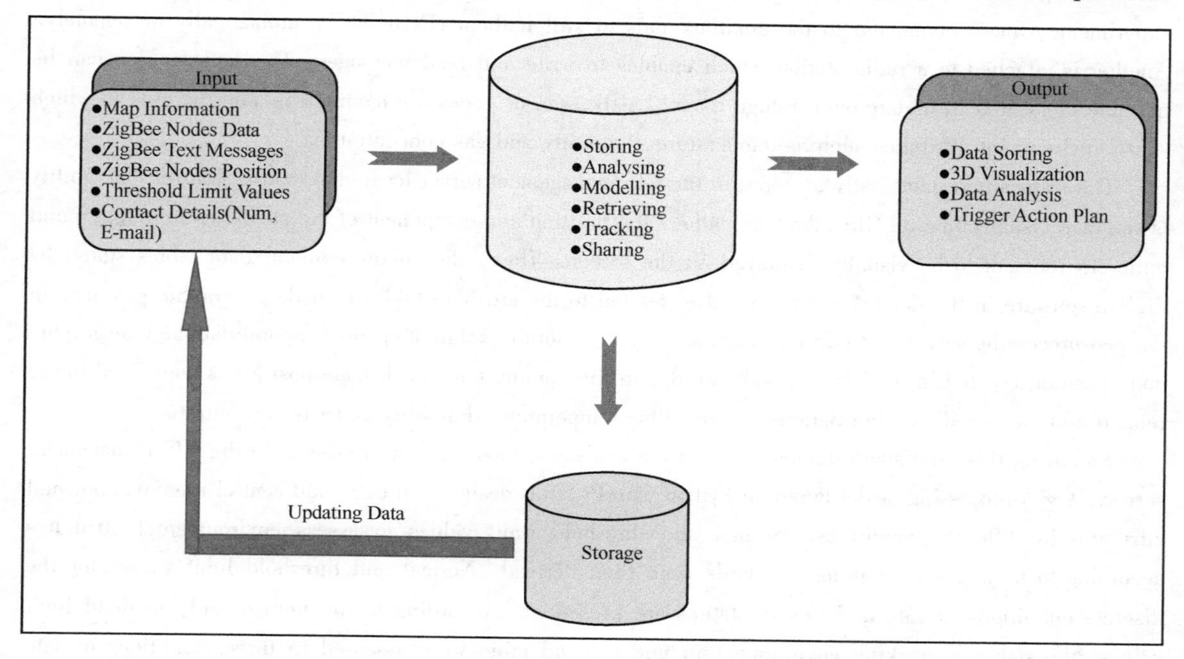

Figure 15-7 Data Flow Sheet of Integrated System in GIS Server

The quality of input dataset is considered to process and analyse at any particular database. Consequently, the quality of input data in our designed GIS management server is divided between long-term and short-term datasets. Maps, ZigBee node positions, threshold limit values and contact details are determined to be long-term input data into the database which may be periodically updated. These data are stored in attribute tables that are associated with ArcGIS geo-processing models. ZigBee node data which measured environmental properties of mine tunnels is derived as short-term (temporal) data. In this case, the datasets of environmental phenomena such as temperature, ***humidity*** and gas concentrations change from time to time or remain relatively continuous. Therefore, spatio-temporal data models which show both spatial and temporal characteristics of environment, are considered as input data in the GIS management server. The spatio-temporal data is stored and manipulated in the ArcGIS geo-processing based on the related or joined table command to digital tables of data collection by ZigBee gateway software.

(2) Process Strategy

Real-time process strategy for safe working environments involves the combination of data models and programs in GIS management server to monitor and communicate underground mine automatically and remotely.

A pattern of decision making in managing spatio-temporal data was modelled as a procedure to monitor the environment attributes of underground mine tunnels. To this end, near real-time and flexible scheduling strategy was planned to apply the performance of ZigBee network in an emergency status. An experiment was simulated on real maps of underground mine with developed ZigBee nodes to control ventilation fans (ON/OFF) and text emergency messages from surface control office. In this model a gateway was located in the surface control office to receive and transmit data through the underground network. The network is extended by ZigBee routers (SRx and UGRx) between the surface gateway and underground end devices based on optimised communication ranges. ZigBee end devices were divided into three groups in this experiment. One is connected to the auxiliary fans to switch them ON or OFF automatically or remotely. Another is attached to a radio station which enables to write and read messages. The radio station can be portable or located in underground refuge room. Lastly, sensor nodes are mounted in working area in which sense environment attributes such as temperature, humidity and gas concentration.

The transmitted data firstly is stored in the GIS management server located at control centre. The ability of the map visualization on GIS (ArcGIS) allows the position and component of the attributes in underground mine environment to be visually displayed on the screen. Then, the spatio-temporal data tables stored by ZigBee software in the database were joined or related to the attribute tables of node geographic positions in the geo-processing services of GIS management server. A joining table of spatio-temporal data and geographic node position created in ArcGIS. In other words, in this joining table each node position is connected to the related and measured variable parameters including temperature, humidity and gas concentration.

Following this, the spatio-temporal data were analysed, modelled and retrieved in the GIS management server. A geo-processing model based on Python (ArcPy) was designed to track and control the environmental attributes in different conditions. Normal and threshold limit values to assess environmental attributes according to underground mining standards were then derived. Normal and threshold limit values for the discrete conditions of safe and unsafe statues are presented. According to the normal and threshold limit values, the status of working environment in underground mine were assessed in three conditions of safe (green), transient (orange) and unsafe (red).

Finally, a loop of conditional procedures and trigger actions were set. The measured parameters (spatio-temporal data) were stored while these data are less than or equal normal limit values (safe condition). The loop was periodically retrieved each 30 min in order to consume less power and to extend the battery life of ZigBee nodes and reduce congestion through the network. Otherwise, a trigger plan was set for the values mounted in the range of between normal and threshold limits (transient condition) or greater than threshold limit (unsafe condition). The trigger action plan applied in the GIS management server to respond the deviation of values from normality. In the transient (orange) condition, the auxiliary fans which had designed for emergency ventilation system would be automatically or remotely turned on. In this state, the model was also setup to send alert messages to shift supervisors. The periodic time of data reading in orange state is reduced to 15 min to ensure the safe and health conditions of underground environment in the shorter time possible. Emergency (alarm) messages in the event of unsafe (red) condition would be texted to surface authorise and to underground personnel for immediate evacuee from the hazardous places. The cycle time of data acquisition is minimised to 5 min in this situation.

(3) Outputs

Mine safety and health were improved by intelligent maps supporting spatio-temporal data and coordinate of ZigBee nodes in this experiment. The final outputs of GIS management server are comprised of 3D visualisation monitoring of underground mine tunnels and messages texting for alert and alarm conditions. The web-GIS is another application supporting the GIS management server to promote the underground monitoring and communication system.

(4) Data Storage

Data storage and management in the central data ***repository*** of server is an essential part of the integrated system. In fact, all geographic and spatial data are stored and managed in ArcMap's geodatabase which accesses to the database at any time over the long term. In the geodatabase, organisational structure for storing datasets and creating relationships between datasets were also provided for further analysis and interpretation. In addition, a multi-user access is enabled to work and command orders from different mine site offices.

An integrated data management and documentation to generate geospatial metadata was another approach of geodatabase automation. Metadata can create geospatial data document to investigate any genuine or non-genuine claims.

New Words and Expressions

underground mine safety and health　　地下矿山安全与健康
prevalence [ˈprev(ə)l(ə)ns]　　*n.* 流行，卓越
ventilation system　　通风系统
bilateral [baɪˈlæt(ə)r(ə)l]　　*adj.* 双方的，双边的，双侧的，(大脑)两半球的
configure [kənˈfɪgə]　　*v.* 使成形，安装
auxiliary [ɔːgˈzɪlɪəri]　　*adj.* 辅助的，备用的
threshold [ˈθreʃəʊld]　　*n.* 阈，门槛，起点，开端
manipulation [məˌnɪpjʊˈleɪʃ(ə)n]　　*n.* 手法，控制，操作法，(熟练的)操作
humidity [hjʊˈmɪdɪti]　　*n.* 湿热，(空气中的)湿度，高温潮湿
repository [rɪˈpɒzɪt(ə)ri]　　*n.* 仓库，贮藏室，存放处，学识渊博的人

Unit Sixteen

16

Product Reliability

Product reliability conveys the concept of dependability, successful operation or performance, and the absence of failures. It is an external property of great interest to both manufacturer and consumer. Unreliability (or lack of reliability) conveys the opposite.

1. Definition of Reliability

The reliability performance of a product is defined as: "The ability of a product to perform required functions, under given environmental and operational conditions and for a stated period of time."

According to this definition, reliability is a rather ***vague*** characterization of the product's ability to perform required functions. Some authors and standards are using the term dependability with the same meaning.

In many applications, the term *reliability* is also used as a measure of the reliability performance, as "the probability that a product will perform one or more specified functions, under given environmental and operational conditions for a stated period of time".

All products degrade with age and/or usage. When the ***product performance*** falls below a desired level, then the product is deemed to have failed. Failures occur in an uncertain manner and are influenced by factors such as design, production, installation, operation, and maintenance. In addition, the human factor is also important.

2. Consequences of Failures

(1) Customer Point of View

When a failure occurs, no matter how benign, its impact is felt. For customers, the consequences of failures may range from mere nuisance value via serious economic loss to something resulting in serious damage to the environment and/or loss of life. All of these lead to customer dissatisfaction with the product.

When the customer is a business enterprise, failures lead to downtimes and affect the production of goods and services. This in turn affects the goodwill of the clients as well as the bottom line of the balance sheet.

(2) Manufacturer Point of View

Lack of reliability affects the manufacturer in a number of different ways. The first is affect on sales due to negative word-of-mouth effects resulting from customer dissatisfaction. This in turn affects the market share and the manufacturer's reputation. The second is higher ***warranty*** costs, resulting from servicing of claims under warranty.

Sometimes, regulatory agencies (e. g., the US Federal Transport Authority) can order the manufacturer to recall a product and replace a component that has not been designed properly from a reliability point of view. In some cases, the manufacturer is required to provide compensation for any damage resulting from failures of the product.

There is no way that a manufacturer can completely avoid product failures. Product reliability is determined by pre-production decisions and impacts the post-production outcomes. The challenge to the manufacturer is to make decisions that achieve a balance between the costs of building-in reliability ***versus*** the consequences of the lack of adequate reliability.

3. Product Safety Requirements

(1) Introduction

Safety is an important attribute of many products. To protect people and the environment, a number of laws, regulations, and standards give requirements to product safety. A main difference between safety requirements and the requirements that have been discussed so far, is that safety requirements are often ***mandatory*** and cannot be traded based on cost-benefit arguments.

Product hazards arise from many different origins. Many product hazards are linked to ***component*** failures. Other hazards stem from the design of the product and may cause injuries and health problems even if the product is without any failure. Examples comprise cutting hazards caused by sharp edges, choking hazards caused by small parts that get loose on children's toys, and health hazards caused by poisonous paint on toys. Product hazards may lead to incidents ranging from the ***nuisance*** category up to disasters.

Some products are installed to protect against major accidents. One such product is the safety instrumented systems. A failure of an SIS will usually not have any significant safety impact on the SIS as such, but may have fatal effects on the system the SIS is protecting, be it a high speed train, a process plant or a nuclear power station.

Product safety requirements are not only stated for the final product, but are related to the whole life cycle of the product, from early conception until disposal. It is often not sufficient to document that the whole life cycle of the product is safe; the manufacturer must also document the activities taken to ensure that the safety is adequate. Detailed risk analyses of the product in the various life cycles are sometimes required, and the risk analysis reports may have to be part of the product ***documentation***.

The safety requirements are strongly dependent on the type and application of the product. To illustrate the nature of the safety requirements and how they are applied, we will use the EU Machinery Directive as a basis for most of the presentation and discussion in this passage.

Product safety requirements are mainly given in laws, regulations, and standards, but may also be stated by the customers of specialized products. Sometimes the requirements may also come from consumer organizations or customer interest groups.

(2) Essential Health and Safety Requirements

This section presents some Essential Health and Safety Requirements (EHSRs) in the EU Machinery Directive. The approach adopted in the Machinery Directive is representative of other product directives in the EU and also for recent legislation in other parts of the world.

A high number of EHSRs are listed in Annex I of the Machinery Directive. The EHSRs in Annex I define the results to be attained, or the risks to be dealt with, but do not specify the technical solutions for doing so. The suppliers are free to choose how the requirements are to be met.

The EHSRs are written in such a way that they remain valid over time, and do not become obsolete with technical progress. Assessment of whether requirements have been met should be based on the state of technical know-how at a given moment. This does not mean that essential requirements are vague. They are drafted in such a way as to give sufficient information to enable assessment of whether or not products meet them.

Some of the requirements are specific and straightforward to fulfil, while other requirements are more complex. This is illustrated by the following sample of requirements:

1) Safety and reliability of control systems: Control systems must be designed and constructed so that they are safe and reliable, in a way that will prevent a dangerous situation arising. Above all they must be designed and constructed in such a way that:

- they can withstand the rigours of normal use and external factors;
- errors in logic do not lead to dangerous situations.

2) Starting: It must be possible to start machinery only by voluntary actuation of a control provided for the purpose.

The same requirement applies:

- when restarting the machinery after a stoppage, whatever the cause;
- when effecting a significant change in the operating conditions;
- unless such restarting or change in operating conditions is without risk to exposed persons.

3) Risks due to falling or ejected objects: Precautions must be taken to prevent risks from falling or ejected objects.

The requirements are mandatory, legally binding obligations, and they are enforced. It is not always possible to meet all the requirements. In this case, the machinery must, as far as possible, be designed and constructed with the purpose of approaching the requirements.

4. EU Directives

An EU Directive is a "law" that is binding for the member states of the EU. A Directive must be transposed into national laws within a specified time ***interval*** after the Directive is issued.

In 1985, a new approach to the development of EU Directives was introduced. The directives that are developed according to this new approach are called New Approach Directives. They present the main requirements for products and systems, while details are left to so-called harmonized standards that are issued by the standardization organizations (e. g. , CEN). A high number of Directives and standards give requirements to product safety. A main objective of these directives is to eliminate differences between national laws, and thereby eliminate trade barriers between the EU Member States. The New Approach Directives provide a basis for technical harmonization. The new scheme is embodied in the regulation on CE Marking, and incorporates conformity assessment procedures directly into the directives.

The Machinery Directive has two main objectives:

- promote free movement of machinery within the EU "single market";
- guarantee a high level of protection to EU workers and citizens.

The Machinery Directive promotes harmonization through a combination of mandatory EHSRs and harmonized standards. The Directive applies to new machinery products that are intended to be placed (or put into service) on the EU market for the first time.

Machinery is defined in the Directive as:

"An assembly of linked parts or components, at least one of which moves, with the appropriate actuators, control and power circuits, etc. , joined together for a specific application, in particular for the processing, treatment, moving or packaging of a material."

The terms "machinery" and "machine" also cover an assembly of machines which, in order to achieve the same end, are arranged and controlled so that they function as an integral whole. This implies that conformance with the EHSRs has to be documented both for single machinery units and for complex systems comprising several machines.

The Machinery Directive is only concerned with injuries to persons and health effects. Consequences to the environment and material/financial assets are not covered in the Directive.

Guidance on how to fulfil the EHSRs and technical solutions for this end are given in harmonized EN-standards. Approximately 750 such EN-standards have been prepared to support the Machinery Directive. If the product complies with the harmonized standards, it is presumed that the product fulfils the EHSRs. It is, however, voluntary to use the standards, and it is, at least in principle, possible to conform to the EHSRs without following the standards, but this will require a rather comprehensive documentation.

(1) The General EU Product Safety Directive

The general EU Product Safety Directive applies to engineering products that can be directly used by consumers. While the safe use of these products is covered by a number of specific technical directives, the Product Safety Directive places additional post-marketing obligations on manufacturers. This directive has also set up a harmonized rapid alert system, called RAPEX, which helps coordinate national market ***surveillance*** against unsafe consumer products.

(2) CE Marking

The CE mark in Figure 16-1 is a mandatory safety mark on many products that are placed on the market in the European Economic Area (EEA).

Figure 16-1 The CE Mark

Officially, CE has no meaning as an ***abbreviation***, but may have originally stood for Communaute Europeenne or Conformite Europeenne, French for European Conformity.

The CE mark is a sign of ***conformity with*** the EHSRs set out in the EU Directives. To permit the use of a CE mark on a product, proof that the item meets the relevant requirements must be documented.

The Machinery Directive splits machinery into two categories: extremely dangerous machines and normal machines. The dangerous machines are listed in Annex IV of the directive. These products are subject to special requirements that involve a Notified Body. A Notified Body is an organization that has been nominated by a member government and notified by the European Commission. The primary role of a Notified Body is to provide services for conformity ***assessment*** related to the conditions set out in the directive in support of CE marking. This normally means assessing the manufacturers conformity to the EHSRs. The conformity assessment can be based on inspection, quality assurance, type examination or design examination, or a combination of these.

For normal machines, the conformity assessment may be done by a company-internal self-certification process.

The responsible organization (manufacturer, representative, and importer) has to issue a Declaration of Conformity indicating his identity (e. g. , name, location), the list of European Directives he declares compliance with, a list of standards the product complies with, and a legally binding signature on behalf of the organization. The Declaration of Conformity underlines the sole responsibility of the manufacturer.

The CE mark is aptly called the passport to Europe for products. All manufacturers, European, American, Chinese, or other, are required to affix the CE mark to products that are governed by New Approach Directives. There are about 25 Directives requiring CE marking.

5. Risk Assessment

To prove conformance with the essential health and safety requirements, risk assessments of the new products may have to be carried out. For some specific types of products, where a C-standard is available, it is possible to claim conformance without a risk assessment. If required, the risk assessment must be carried out according to the standard ISO 14121-1.

The main steps of the risk assessment are:

1) Define machine: This step involves describing the machine, its intended use, space and time limits, and boundaries and interfaces for all life cycle phases. The life cycle phases of a machine are, according to ISO 12100-1:

- construction;
- transport, assembly and installation;
- commissioning;
- use:

-setting, teaching/programming or process changeover;

-operation;

-cleaning;

-fault finding;

-maintenance;

- de-commissioning, dismantling and, as far as safety is concerned, disposal.

2) Identify hazards: Here, all hazards and hazardous situations considering the various aspects of the operator-system relationship, the possible states of the machine and reasonably foreseeable misuse must be identified. Hazards can be classified as continuing hazards, which are inherent in the machine, material or substance; and hazardous events that can result from machine failures and human errors. The concept of

reasonably foreseeable misuse is important. It is not sufficient that the product/machine is safe during its intended use conditions. It must also be safe for foreseeable misuse conditions. These include situations where operators make shortcuts and use simplified operating procedures, and where other people use the machine for other purposes, for example, children who play with the machine.

The hazard identification may be done as a preliminary hazard analysis, an FMECA or an HAZOP. For more complex systems, it may also be necessary to use methods like fault tree analysis and event tree analysis. The analysis may be supported by the list of generic hazards in Annex A of ISO 14121-1.

3) Analyse consequences: In this step, the consequences or harm related to potential incidents are identified and analysed. These primarily relate to injury and ill health as a result of exposure to a hazard. It can also be described in terms of economic losses due to interruption to production and asset damage or in terms of environmental damage. The severity of the consequences must be assessed based on some predefined scale.

4) Estimate risk: Risk is generally defined as a function of chance (probability) of the harm being realized and the consequences (severity) of this harm. In ISO 14121-1 risk is a function of:

① The severity of harm.

② The probability of occurrence of that harm, which is a function of:

- the exposure of person (s) to the hazard;
- the occurrence of a hazardous event;
- the technical and human possibilities to avoid or limit the harm.

The elements of risk are illustrated in Figure 16-2.

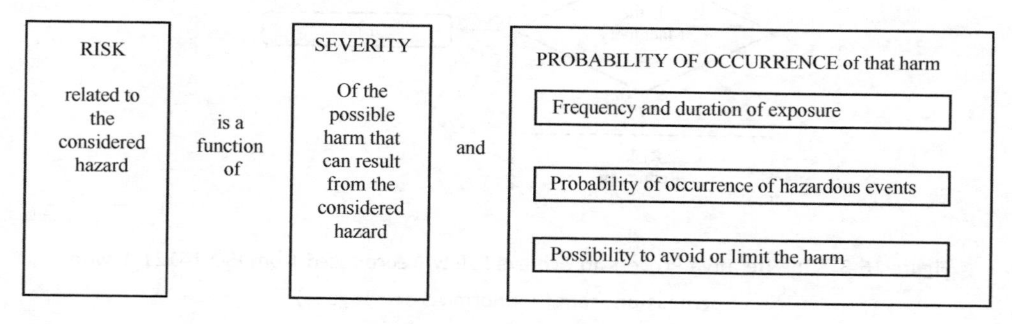

Figure 16-2 The Elements of Risk

5) Evaluate the risk: The risk is evaluated based on some predefined criteria. The purpose of this step is to decide if risk is tolerable or should warrant some corrective or preventive measures.

6) Risk control strategy: If risk is judged to be tolerable, a hierarchy of risk reducing options is set out in the Machinery Directive. In selecting the most appropriate methods, the manufacturer must apply the following principles, in the order given:

- eliminate or reduce risks as far as possible (inherently safe design and construction), see also Kivistö-Rahnasto (2000);
- take the necessary protection measures in relation to risks that cannot be eliminated, e. g. , see Kjellén (2000);
- inform the users about the ***residual*** risks due to any shortcomings of the protection measures adopted, indicate whether any particular training is required and specify any need to provide personal

protection equipment.

7) Verification: There will be a need to review the system following modifications to ensure that these measures will reduce risks to a tolerable level and that no new hazards are generated as a result of design changes.

The iterative process to achieve safety as outlined in ISO 14121-1 is illustrated in Figure 16-3.

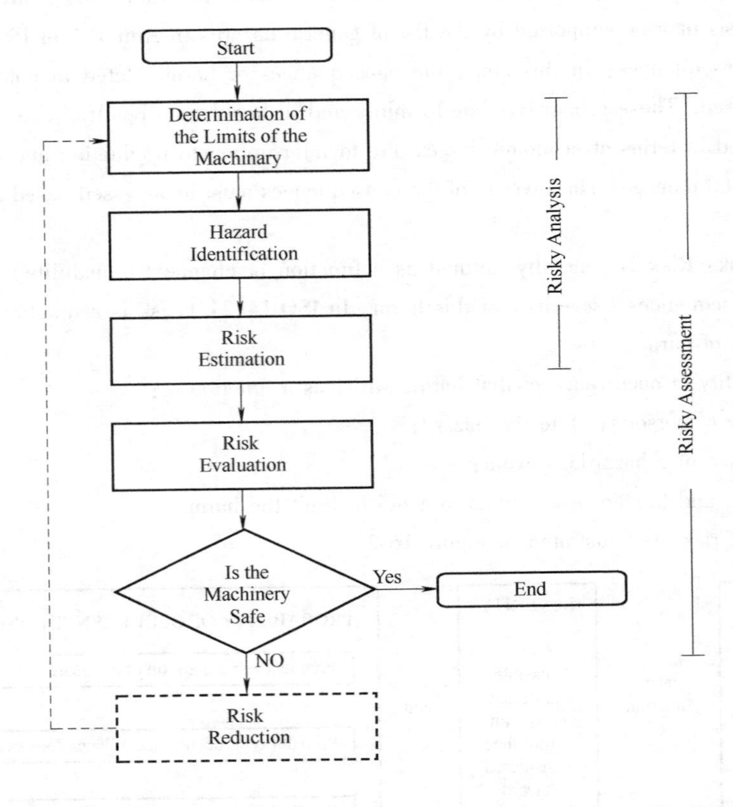

Figure 16-3 The Iterative Process to Achieve Safety (reproduced from ISO 14121-1 with permission from Pronnorm AS-see page ix)

New Words and Expressions

vague [veɪg] *adj.* 含糊的，不明确的，不清楚的，含糊其辞的

product performance 产品性能

warranty [ˈwɒr(ə)nti] *n.*（商品）保用单

versus [ˈvɜːsəs] *prep.* 与，对

mandatory [ˈmændət(ə)ri] *adj.* 强制的，法定的，义务的

component [kəmˈpəʊnənt] *n.* 成分，部件，组成部分

nuisance [ˈnjuːs(ə)ns] *n.* 麻烦事，讨厌的人（或东西），妨害行为

documentation [ˌdɒkjʊmenˈteɪʃ(ə)n] *n.* 证明文件，归档，必备资料，文件记载

interval [ˈɪntəv(ə)l] *n.* 间歇，音程，休息时间

surveillance [səˈveɪl(ə)ns] *n.* 监视，监督

abbreviation [əbriːvɪ'eɪʃ(ə)n]　　n. 缩写，约分，缩略语，〔数〕约分
conformity with　　与……一致
assessment [ə'sesmənt]　　n. 评价，评定，判定，看法
residual [rɪ'zɪdjʊəl]　　adj. 剩余的，残留的

Exercises

1. Explain the definition of reliability.
2. What are consequences of failures from customer's and manufacturer's point of view?
3. What are requirements for product safety? Give examples of Essential Health and Safety Requirements which are specific and straightforward to fulfil.
4. Explain EU Directives and find some EU Directives in your life or website.
5. What are the main steps of the risk assessment?

科技英语翻译技巧——as 的用法

在科技英语中，as 的用途和含义有：连词、介词、关系代词、as 构成的成语和词组。

1. as 用作连词

as 用作连词，可引导时间状语从句、原因状语从句、方式状语从句、比较状语从句、让步状语从句。

（1）as 引导时间状语从句

as 引导的时间状语从句，表示主句与从句是在同一时间发生的，一般可以为“当……时”“一……就”“随着……”等。例如：

The antenna gain becomes larger as frequency is increased in the microwave band.

在微波波段，天线增益随着频率的增大而增大。

As soon as the force producing motion was removed the body would stop.

产生运动的力一旦除去，物体就会停止运动。

（2）as 引导原因状语从句

as 引导原因状语从句时，可翻译成“由于”“因为”“既然”等。例如：

Electrical energy is widely used in everyday life as it can be easily transformed into light and heat.

由于电能易于转化为光和热，所以在日常生活中得到广泛使用。

As liquids and gases flow, they are called fluids.

由于液体和气体能流动，因此称为流体。

（3）as 引导方式状语从句

as 引导方式状语从句时，译为“像……一样”“正如……”“似乎”等。例如：

Just as there is always a powerful attraction between two unlike electric charges, so there is a powerful repulsive force between two like electric charges.

正如两种异性电荷间总有强大的引力一样，两种同性电荷间也存在强大的斥力。

Light waves are different in frequency as sound waves are.

光波和声波一样有不同的频率。

（4）as 引导比较状语从句

as 引导比较状语从句的形式为：

1）as + 形容词或副词 + as，可译为“与……一样”。

2）not so + 形容词或副词 + as，可译为“不如（没有）……（那样）”。

3）as long as，可译为“多久……”。

例如：

Heat is another form of energy, which is as important as other kind of energy.

热是能量的另一种形式，与其他形式的能量同样重要。

（5）as 引导让步状语从句

as 引导让步状语从句的结构为：“形容词 + as + 主语 + 系词”，译为“虽然……但……”。例如：

Small as atoms are, electrons are still small.

原子虽然很小，但电子更小。

2. as 用作介词

as 用作介词时，可引出同位语、补足语，可翻译为“作为”“为”“以”等。

（1）as 引出同位语

Heat as a physical quantity must be measured in terms of a definite unit.

热作为一个物理量，必须用一定的单位来度量。

（2）as 引出补足语

在科技英语中，as 可引出宾语补足语和主语补足语，常见的形式有：

accept. . . as. . . ，可译为“把……作为……，认为……是……”。

classify. . . as. . . ，可译为“把……分为……”。

regard. . . as. . . ，可译为“把……称作……”。

define. . . as. . . ，可译为“把……定义为……”。

例如：

We can define an electric current as a controlled motion of electrons.

我们可以把电流定义为电子的受控流动。

Maxwell’s equations simplify into a subject commonly referred as geometrical optics, which treats light as a ray traveling on a straight line.

麦克斯韦方程简化为通常所说的几何光学，他将光视作在直线上传播的光线。

3. as 用作关系代词

as 用作关系代词可引导定语从句，可以单独使用，也可以与 such、the same 等词语搭配使用。

（1）as 单独使用

As the same indicates, a semiconductor material has poorer conductivity than a conductor, but better conductivity than an insulator.

顾名思义，半导体材料的导电性比导体差，但比绝缘体好。

（2）the same. . . as 引出定语从句　可译为“和……一样”“与……相同”。例如：

The current is in the same direction as the motion of the positive particles.

电流的方向即正电荷运动的方向。

（3）such. . . as 引导定语从句　可译为“像……这样的”“像……之类的”。

The motion of ions is the motion of such atoms as have gained or lost electrons, which in most cases takes place in chemical solutions.

离子的运动也就是原子得到或失去电子时的运动，在大多数情况下，化学溶液中发生这种现象。

Human Factors Contributions to Consumer Product Safety

Human Factors (***HF***) refers to designing products and systems for human use. As Don Norman, leader in human-centered design applications, has stated, human interaction with any product is heavily influenced by the users' goals and interpretations of the product. At the CPSC (Consumer Product Safety Commission), HF involves studying the user, product characteristics and use, and environment separately, and in combination, determining the factors that contributed to an injury-producing situation. HF staff evaluates the consumer and the product as a system—in other words, how human behavior affects the safety of the product, and how the design of the product affects human behavior. Human behavior includes all the aspects of the person—from their physical attributes and how that affects their interaction with the product—to information processing, reaction times, and expectations.

1. CPSC History and Background

Congress established the CPSC in 1972, as an independent federal health and safety regulatory agency, charged with protecting the public from unreasonable risks of injury or death associated with the use of the thousands of consumer products under the agency's ***jurisdiction***. These products are used in the home, schools, in recreation, other public places, or otherwise. Based on estimates from the CPSC's Directorate for Economic Analysis, deaths, injuries and property damage from consumer product incidents cost the nation more than $1 trillion annually. The CPSC is committed to protecting consumers and families from product-related hazards that pose an unreasonable risk of injury. The CPSC achieves that goal through education, safety standards activities, regulation, and enforcement of the statutes and implementing regulations.

As a result, the CPSC is charged, specifically, with the following:

1) Protecting the public against unreasonable risks of injury associated with consumer products.

2) Assisting consumers in evaluating the comparative safety of consumer products.

3) Developing uniform safety standards for consumer products and to minimize conflicting state and local regulations.

4) Promoting research and investigation into the causes and prevention of product-related deaths, illnesses, and injuries.

The CPSC's authority generally requires the agency to rely on voluntary standards rather than ***promulgate*** mandatory standards, if compliance with a voluntary standard would eliminate or adequately reduce the risk of injury identified, and if it is likely there will be substantial compliance with the voluntary standard. CPSC staff works with both domestic and international standards development organizations that coordinate the development of voluntary standards.

Mandatory standards, on the other hand, are federal rules set by statute or regulation that impose requirements for consumer products. Mandatory standards under the ***Consumer Product Safety Act*** (***CPSA***) typically take the form of performance requirements that consumer products must meet or warnings they must display to be imported, distributed, or sold in the United States.

Under the CPSA, the CPSC may set a mandatory standard when it determines that ***compliance*** with a voluntary standard would not eliminate or adequately reduce the risk of injury or finds that it is unlikely that there will be substantial compliance with a voluntary standard. Under the CPSA, the Commission may also promulgate a mandatory ban of a hazardous product when the Commission determines that no feasible mandatory standard would adequately protect the public from an unreasonable risk of injury. In some cases, Congress directs and authorizes the Commission to promulgate a mandatory standard.

2. Human Factors Analysis in Consumer Product Safety

HF staff works together with our partner divisions in Engineering Sciences, Health Sciences, and Laboratory Sciences to develop suggestions for product redesign after a product has been manufactured. Human factors studies and research applies the psychology of human behavior, learning theory, engineering principles, anthropometry, biomechanics, and industrial design to consumer product evaluation. This total system evaluation includes the person + the product + the environment. The analysis and testing conducted by HF staff directly contributes to consumer product safety and risk reduction through design verification and validation efforts. Specifically, through technical analyses and reports, HF staff addresses:

- what age range of children is most likely to use the product if the product is intended for use primarily by children;
- how consumers interact with and use the product, such as foreseeable use and misuse;
- the likely effectiveness of alternative designs, guarding systems, warnings, labels, and other hazard-mitigation strategies;
- the likelihood of consumers encountering a given hazardous scenario;
- anthropometric and strength evaluation (i. e., the size of the person and their strength capabilities);
- the ability of consumers to perform a repair or carry out a desired corrective action;
- how consumer behavior and product design interact.

At CPSC, human factors involve studying the interaction of humans, products, and the environment, the parameters of human performance, the relevant design features of consumer products, and how all of these elements integrate and contribute to a potentially hazardous situation.

(1) Origin of Human Factors Work

Human factors work ***comes to*** the CPSC through various ***avenues***. This includes, consumers reporting incidents through the agency's hotline or website; field investigators conducting interviews with victims and their families and compiling photographs and details from police reports; port surveillance personnel who send potentially violative products that are intended for import into the United States from various countries; news reports of consumer product-related incidents; death certificates and medical examiner reports; and from manufacturers themselves who have a legal obligation to report to the CPSC. This legal obligation is specified under the CPSA, requires manufacturers to report on a product that: ① is defective, ② does not comply with standards and regulations, ③ was choked on, ④ has been specified as a substantial product hazard by CPSC, or ⑤ was subject to certain lawsuits. Human factors analysis is also part of all project work and ***petitions*** that are examined at the agency.

(2) Age Determinations

To enforce CPSC regulations, it is often necessary to determine the intended ages of the users of toys, products, and jewelry. These age determinations are needed to assess products for defects, evaluate a corrective action plan, and conduct enforcement activities. Currently, determinations are made on a toy-by-toy basis through ***consultation*** with child development resources. Attention is given to the developmental abilities of typical children, with an understanding that a wide variation exists within the rates at which children achieve major developmental ***milestones***.

(3) Toy Evaluations

Toy evaluations are conducted by laboratory engineers via tests that simulate the use and foreseeable abuse of toys by children. A toy must pass all of the applicable standards tests and are required to be certified by third party testing laboratories as complying with those standards. The CPSC National Product Testing and Evaluation Center (NPTEC) conducts either spot-checks of the compliance or evaluations based on an incident to determine if the product can continue to be sold to the public. Some examples include: drop tests, torque tests, pull tests, small-parts tests, and sharp-edge tests. Specific forces, sizes of clamps, and distances of drops have been determined to mimic as closely as possible those produced by a child during normal play situations and foreseeable misuse.

(4) Product Safety Assessments

The CPSC's Office of Compliance negotiates recalls, liaises with companies, and encourages manufacturers to comply with CPSC regulations. A PSA, Product Safety Assessment, is the CPSC-generated document that provides the scientific and technical analysis conducted by technical staff on that specific product. Most PSAs are associated with particular hazards that human factors will focus on; and in general, only hazards that have been associated with known incidents are addressed. For example, if incidents resulted from a fall, HF staff evaluates the potential fall hazards associated with that product sample.

Often, if HF staff is asked to assess a corrective action plan that involves a product repair, the usability analysis involves the question: "Could the average consumer successfully complete the repair?" In these cases, HF staff may conduct several different kinds of analyses. For example, staff would provide a usability review and make recommendations regarding manufacturer-provided instructions and also conduct a walkthrough of the recommended repair to determine whether the repair can be completed successfully based on the materials provided, the instructions, and the likely abilities of the intended user.

Often, the CPSC HF staff is asked to perform usability analyses on instructions, which might include evaluating warnings and labels. Generally, the CPSC HF staff recommends warnings adhere to the ANSI Z535 series of standards, which are the primary US voluntary consensus standards for safety signs and colors. Figure 16-4 presents an example of a label that staff evaluated, which was determined not to have followed good warning practices. Although the label adheres to ANSI standard warning color use, the average US consumer cannot understand the last bullet because "oversely" is not a word. In addition, Hilts and Krilyk (1991) reported that adults read at least one or two grade levels below their last school grade completed, and the Flesch-Kincaid Readability Calculator estimates the language in this label at the 14th grade level.

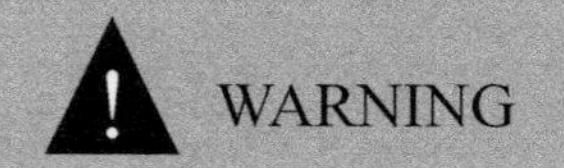

Figure 16-4 Example Warning Label Evaluated by Human Factors Staff at the CPSC as Part of a PSA

(5) Projects and Petitions

PSAs examine hazards associated with specific models of a product, manufactured by a particular firm; whereas, projects and petitions examine hazards and hazard patterns associated with whole classes or categories of products. So while a PSA request might ask staff to analyze model X of a particular firm's line of clothes dryers, a project on clothes dryers may ask for an analysis of hazards that are common to most or all clothes dryers currently on the market. Projects and petitions generally take place over a time frame of several months or years. The results of these analyses may be used to develop or modify consumer product safety standards, which would affect all future production of that product.

HF staff does a large amount of project work related to children's products because of the requirements in the Consumer Product Safety Improvement Act (CPSIA) of 2008. Section 104 of the Act requires the CPSC to study and develop safety standards for ***infant*** and toddler products. The CPSC's standard must be substantially the same as the relevant voluntary safety standard or more ***stringent*** than that voluntary safety standard. One example is the Children's Portable Bed Rails project. Portable bed rails are products with a vertical panel that rests against the side of a bed to prevent children from falling out of adult-sized beds. Portable bed rails typically have support legs that slide between the mattress and box spring. However, earlier versions of these bed rails can sometimes slip out of place and allow a child to slide between the rails and the bed. The child who becomes entrapped in this space can ***suffocate***. HF staff was asked to design a probe to be incorporated into a standard and used during performance testing of bed rails. This probe was designed based on the anthropometric dimensions of the children at risk, and all portable bed rails are now required to prevent passage of the probe when a certain force is applied.

While projects and petitions are similar in scope, there are some differences between the two. Petitions are essentially requests made by the public for the CPSC to issue, amend, or revoke a rule or regulation associated with a particular product or product hazard. Typically, work performed in response to a petition typically is completed within 180 days; accordingly, technical assessments must be made using existing or "off-the-shelf" data. New studies or data collection are generally not considered.

In human factors, analysis might include incident analysis, and specifically, analysis of behavioural and anthropometric contributing factors. We could also evaluate provided instructions and warnings and recommend design modifications that increase safety and enhance human performance.

3. Methodology

HF staff at the CPSC uses a wide variety of methodologies and approaches to analyze product safety and design, from ***heuristics*** to task analysis. The human factors engineers and psychologists have the autonomy to determine which approach is right for the product on hand and will work best to aid the mechanical engineers, health scientists, and Office of Compliance personnel to make a determination on how to proceed in interacting with the manufacturer.

(1) Heuristics

Heuristics require the Human Factors analysts to use professional judgement, experience, and best practice education to guide product evaluation (Nielsen, 1992). Heuristic analysis at the CPSC is conducted in both table-top and simulated settings at the CPSC National Product Testing and Evaluation Center (NPTEC). While this is a subjective analysis, conducted by subject matter experts, the approach is often a preferred method for completing analysis, due to the ease and speed which work at the agency often demands.

(2) Surveys

Standardized questionnaires and surveys have been used by staff to assist in making informed and accurate expert assessments of consumer behavior. The user population that the CPSC supports ranges quite broadly from infants to senior citizens. Therefore, it is often advantageous to conduct a targeted survey to determine how the general public interacts with a specific product design. This methodology is often time-consuming and extensive, requiring funding and approval by the US Office of Management and Budget. Thus, the methodology is used primarily to support long-term projects.

(3) Task Analysis

Task analysis refers to the study of what a user is required to do, in terms of actions and cognitive processes, to achieve a goal presented by a product or system. Staff employs a variety of task analysis techniques to identify how incidents and accidents occur when there is interaction between a product and user. These techniques include Observational Analysis, Task Decomposition, Walk-throughs, Fault Tree Analysis, and on occasion, Computer Modelling and Simulation.

HF task analyses at the CPSC are used to:

- identify hazards to the user;
- analyze product safety through good design techniques;
- analyze human error when interacting with the product and environment; and
- identify remedial measures.

The Division of Human Factors at the CPSC provides support for enforcement and safety initiatives, by evaluating the behavioral interactions between consumers and the products they use. HF analysis involves studying the human interaction, product characteristics and use, and the environment, separately, and in combination, to determine the factors that contributed to a hazardous situation. This technical discipline is an important contributor to the analysis that the CPSC uses to understand and determine the interaction of people with consumer products to reduce the risk of injury and death.

New Words and Expressions

Human Factors (HF) 人性因素，人为因素
jurisdiction [ˌdʒʊərɪsˈdɪkʃ(ə)n] *n.* 管辖权，司法权，审判权，管辖范围
promulgate [ˈprɒm(ə)lgeɪt] *v.* 传播，颁布，传扬，宣传
Consumer Product Safety Act (CPSA) (美) 消费产品安全法
compliance [kəmˈplaɪəns] *n.* 遵从，顺从，服从
come to 谈到，达到，共计
avenue [ˈæv(ə)njuː] *n.* 途径，〈美〉(南北向) 街道，林荫路，道路
petition [pɪˈtɪʃ(ə)n] *n.* 请愿，请愿书，申请，祈求
consultation [kɒnsəlˈteɪʃ(ə)n] *n.* 咨询，磋商，查阅，商讨
milestone [ˈmaɪlstəʊn] *n.* 里程碑，(一生中或历史上的) 划时代事件
infant [ˈɪnf(ə)nt] *n.* 婴儿，幼儿，四岁到七岁的学童
stringent [ˈstrɪn(d)ʒ(ə)nt] *adj.* 严格的，严厉的，紧缩的，短缺的
suffocate [ˈsʌfəkeɪt] *v.* 窒息而死，闷死，让人感觉闷热，憋气
heuristics [hjʊəˈrɪstɪks] *n.* 探索法，启发式

Unit Seventeen

17

Emergency Management

1. Introduction to Disasters and Emergency Management

A disaster is a state in which a population, population group, or an individual is unable to cope with or overcome the ***adverse*** effects of an extreme event without outside help. The impact of an extreme event may include significant physical damage or ***destruction***, loss of life, or drastic change to the environment. It is a phenomenon that can cause damage to life and property and destroy the economic, social, and cultural life of people.

The above definition perceives disasters as the consequence of inappropriately managed risk. For example, when looking at an extreme event such as a flood, although the primary cause for a flood is extreme rainfall, snowmelt, or a combination of both, the impact or magnitude of a flood is determined by human influences. The risks associated with a disaster are a function of both the hazards and the vulnerability of the affected group. Hazards that strike in areas with low ***vulnerability*** will never become disasters, as is the case with uninhabited regions. Developing countries suffer the greatest costs when a disaster hits; more than 95% of all deaths caused by disasters occur in developing countries, and losses due to natural disasters are 20 times greater as a percentage of Gross Domestic Product (GDP) in developing countries than in industrialized countries.

Disaster management is defined as the organization and management of resources and responsibilities for dealing with all humanitarian aspects of emergencies, in particular preparedness, ***mitigation***, response, and recovery in order to lessen the impact of disasters. It deals specifically with the processes used to protect populations and/or organizations from the consequences of disasters. However, it does not necessarily extend to the prevention or elimination of the threats themselves, although the study and prediction of threats is an imminent part of disaster management. International organizations focus on community-based disaster preparedness, which assists communities to reduce their vulnerability to disasters and strengthen their coping capacities.

When the capacity of a community or country to respond to and recover from a disaster is exceeded, outside help is necessary. This assistance comes from different sources including government and nongovernment organizations. The definition of disaster as a state where people at risk can no longer help themselves conforms to the modern view of a disaster as a social event, where the people at risk are vulnerable to the effects of an extreme event because of their social conditions. According to this view, disaster management is not only a technical task, but also a social task. Therefore, community-based disaster preparedness becomes an integral part of disaster management. By reducing a community's vulnerability to a disaster and by building upon its coping capacities, skills, and resources, these same communities are better able to meet future crises.

The first people to respond to a disaster are those living in the local community, and they are also the first to start rescue and relief operations. They know what their needs are, have an intimate knowledge of the area, and may have experienced similar events in the past. Therefore, community members need to be consulted and involved in the response and recovery operations, including ***assessment***, planning, and ***implementation***. Consultation can take place through community leaders, representatives of women's or other community associations, beneficiaries, and other groups.

2. Types of Disasters

Disasters can be classified into two subcategories: natural hazards and technological or man-made hazards. Natural hazards are naturally occurring physical phenomena caused by either rapid or slow onset events that can be geophysical, hydrological, climatological, meteorological, and/or biological in nature. Geophysical disasters include earthquakes, landslides, tsunamis, and volcanic activity; hydrological disasters include hazards such as avalanches and floods; climatological disasters include hazards such as extreme temperatures, droughts, and wildfires; meteorological disasters include cyclones, hurricanes, storms, and/or wave surges; and biological disasters include disease epidemics and infestations such as insect and/or animal plagues. Some natural disasters can result from a combination of different hazards, for example, floods can be the result of tsunamis, storm surges, hurricanes, or cyclones or a combination of all four. However, after a flood, ***epidemics*** such as cholera, malaria, and dengue fever begin to emerge.

Technological or man-made hazards are events that are caused by humans and occur in or close ***proximity to*** human settlements. This can include environmental degradation, pollution, complex emergencies and/or conflicts, cyber-attacks, famine, displaced populations, industrial accidents, and transport accidents. Workplace fires are more common and can cause significant property damage and loss of life. Communities are also vulnerable to threats posed by extremist groups who use violence against both people and property. High-risk targets include military and civilian government facilities, international airports, large cities, and high-profile landmarks. Cyberterrorism involves attacks against computers and networks done to intimidate or coerce a government or its people for political or social objectives.

Technological disasters are complex by their very nature and could include a combination of both natural and man-made hazards. In addition, there is a range of challenges such as climate change, unplanned urbanization, underdevelopment and poverty, as well as the threat of pandemics that will shape disaster management in the future. These aggravating factors will result in the increased frequency, complexity, and severity of disasters.

Climate change ranks among the greatest global problems of the 21st century, and the scientific evidence on climate change is stronger than ever. For example, the Intergovernmental Panel on Climate Change (IPCC) released its Fourth Assessment Report in early 2007, saying that climate change is now unequivocal. It confirms that extremes are on the rise and that the most vulnerable people, particularly in developing countries, face the brunt of impacts. The gradual expected temperature rise may seem limited, with a likely range from 20℃ to 40℃ predicted for the coming century. However, a slightly higher temperature is only an indicator that is much more skewed. Along with the rising temperature, we can experience an increase in both frequency and intensity of extreme weather events such as prolonged droughts, floods, landslides, heat waves, and more intense storms; the spreading of insect-borne diseases such as malaria and dengue to new places where people are less immune to them; a decrease in crop yields in some areas due to extreme droughts or downpours and changes in timing and reliability of rainfall seasons; a global sea level rise of several centimeters per decade, which will affect coastal flooding, water supplies, tourism, and fisheries, and tens of millions of people will be forced to move inland; and melting glaciers, leading to water supply shortages. Climate change is here to stay and will accelerate. Although climate change is a global issue with impacts all over the world, those people with the least resources have the least capacity to adapt and therefore are the most vulnerable. Developing countries, more specifically its poorest inhabitants, do not have the means to cope with floods and other natural disasters; to make matters worse, their economies tend to be based on climate and/or weather-sensitive sectors such as agriculture and fisheries, which makes them all the more vulnerable.

The impact of underdevelopment, unplanned urbanization, and climate change is present in our everyday work; disasters are a development and humanitarian concern. A considerable incentive for rethinking disaster risk as an integral part of the development process comes from the aim of achieving the goals laid out in the Millennium Declaration. The Declaration sets forth a road map for human development supported by 191 nations. Eight Millennium Development Goals (MDGs) were agreed upon in 2000, which in turn have been broken down into 18 targets with 48 indicators for progress. Most goals are set for achievement by 2015. The MDGs have provided a focus for development efforts globally. While poverty has fallen and social indicators have improved, most countries will not meet the MDGs by 2015, and the existing gap between the rich and the poor will widen. Recently, the campaigns on poverty have resulted in key milestones on aid and debt relief. While positive, much more is needed if the MDGs are to become reality. These efforts to reduce poverty are vital for vulnerability reduction and strengthened resilience of communities to disasters.

Today 50% of the world's population live in urban centers, and by 2030 this is expected to increase to 60%. The majority of the largest cities, known as megacities, are in developing countries, where 90% of the population growth is urban in nature. Migration from rural to urban areas is often trigged by repeated natural disasters and the lack of livelihood opportunities. However, at the same time many megacities are built in areas where there is a heightened risk for earthquakes, floods, landslides, and other natural disasters. Many people living in large urban centers such as "slums" lack access to improved water, sanitation, security of tenure, durability of housing, and sufficient living area. This lack of access to basic services and livelihood leads to increasing risk of discrimination, social exclusion, and ultimately violence.

3. Emergency Planning

The terrorist attacks of September 11, 2001, illustrated the need for all levels of government, the private sector, and nongovernmental agencies to prepare for, protect against, respond to, and recover from a wide spectrum of events that exceed the capabilities of any single entity. These events require a unified and coordinated approach to planning and emergency management.

Knowing how to plan for disasters is critical in emergency management. Planning can make the difference in mitigating against the effects of a disaster, including saving lives and protecting property and helping a community recover more quickly from a disaster. Developing an effective Emergency Operations Plan (EOP) can have certain benefits for a municipality including the successful evacuation of its citizens and the ability to survive on its own without outside assistance for several days. The consequences of not having an emergency plan include the following: The need for immediate assistance and a higher number of casualties resulting from an attempted evacuation, with them being ineligible for the full amount of aid from upper-tier municipalities (regions or counties), provincial, state, or federal governments. In the United States, counties that do not have emergency plans cannot declare an emergency and are ineligible for any aid or for the full amount of aid. In Ontario, Canada, all municipalities must have an emergency plan in accordance with the Emergency Management and ***Civil Protection Act***.

Emergency planning is not a one-time event. Rather, it is a continual cycle of planning, training, exercising, and revision that takes place throughout the five phases of the emergency management cycle: preparedness, prevention, response, recovery, and mitigation. The planning process does have one purpose: the development and maintenance of an up-to-date EOP. An EOP can be defined as a document describing how citizens and property will be protected in a disaster or emergency. Although the emergency planning process is cyclic, EOP development has a definite starting point. There are four steps in the emergency planning process: the hazard analysis, the development of an EOP, testing the plan through a series of training exercises, and plan maintenance and revision.

Step 1: The hazard analysis. It is the process by which hazards that threaten the community are identified, researched, and ranked according to the risks they pose and the areas and ***infrastructure*** that are vulnerable to damage from an event involving the hazards. The outcome of this step is a written hazard analysis that quantifies the overall risk to the community from each hazard. The hazard analysis component of the emergency planning process.

Step 2: The development of an EOP. The EOP includes the basic plan, functional annexes, hazard-specific appendices, and implementing instructions. The outcome of this step is a completed plan, which is ready to be trained, exercised, and revised based on lessons learned from the exercises.

Step 3: Testing the plan through a series of training exercises. Training exercises of different types and varying complexity allow emergency managers to see what in the plan is unclear and what does not work. The outcomes of this step are lessons learned about weaknesses in the plan that can then be addressed in Step 4.

Step 4: Plan ***maintenance*** and revision. The outcome of this step is a revised EOP, based on current needs and resources, which may have changed since the development of the original EOP. After the EOP is developed, Step 3 and Step 4 repeat in a continual cycle to keep the plan up to date. If you become aware that your community faces a new threat such as terrorism, the planning team will need to revisit Step 1 and

Step 2. Emergency planning is a team effort because disaster response requires coordination between many community agencies and organizations and different levels of government. Furthermore, different types of emergencies require different kinds of expertise and response capabilities. Thus, the first step in emergency planning is identification of all of the parties that should be involved.

Obviously, the specific individuals and organizations involved in response to an emergency will depend on the type of disaster. Law enforcement will probably have a role to play in most events, as will fire, Emergency Medical Services (EMS), voluntary agencies (i. e., the Red Cross), and the media. On the other hand, hazardous materials personnel may or may not be involved in a given incident but should be involved in the planning process because they have specialized expertise that may be called on.

Getting all stakeholders to take an active interest in emergency planning can be a daunting task. To schedule meetings with so many participants may be even more difficult. It is critical, however, to have everyone's participation in the planning process and to have them feel ownership in the plan by involving them from the beginning. Also, their expertise and knowledge of their organizations' resources are crucial to developing an accurate plan that considers the entire community's needs and the resources that could be made available in an emergency.

It is definitely in the community's best interest to have the active participation of all stakeholders. The following are recommendations of what can be done to ensure that all stakeholders that should participate in the discussions do, so that a plan is formulated. Give the planning team plenty of notice of where and when the planning meeting will be held. If time permits, you might even survey the team members to find the time and place that will work for them. Provide information about team expectations ahead of time. Explain why participating on the planning team is important to the participants' agency and to the community itself. Show the participants how they will contribute to a more effective emergency response. Ask the Chief Administrative Officer (CAO) or their Chief of Staff to sign the meeting announcement. A directive from the executive office will carry the authority of the CAO and send a clear signal that the participants are expected to attend and that emergency planning is important to the community. Allow for flexibility in scheduling after the first meeting. Not all team members will need to attend all meetings. Some of the work can be completed by task forces or subcommittees. Where this is the case, gain concurrence on time frames and milestones but let the subcommittee members determine when it is most convenient to meet. In addition, emergency managers may wish to speak with their colleagues from adjacent municipalities to gain their ideas and inputs on how to gain and maintain interest in the planning process.

Working with personnel from other agencies and organizations requires collaboration. Collaboration is the process by which people work together as a team toward a common goal, in this case, the development of a community EOP. Successful collaboration requires a commitment to participate in shared decision making; a willingness to share information, resources, and tasks; and a professional sense of respect for individual team members. Collaboration can be made difficult by differences among agencies and organizations in terminology, experience, mission, and culture. It requires the flexibility from team members to reach an agreement on common terms and priorities and humility to learn from others' ways of doing things. Also, collaboration among the planning team members benefits the community by strengthening the overall response to the disaster. For example, collaboration can eliminate duplication of services, resulting in a more efficient response and expanding resource availability. It can further enhance problem solving through cross-pollination of ideas.

New Words and Expressions

adverse [ˈædvɜːs]	*adj.* 不利的，有害的，反面的
destruction [dɪˈstrʌkʃ（ə）n]	*n.* 摧毁，破坏，毁灭，消灭，灭亡，扑灭
vulnerability [ˌvʌlnərəˈbɪləti]	*n.* 脆弱性，弱点，易伤性，可捕性
mitigation [mɪtɪˈgeɪʃ（ə）n]	*n.* 减轻，缓和，平静
assessment [əˈsesmənt]	*n.* 评定，估价
implementation [ɪmplɪmenˈteɪʃ（ə）n]	*n.* ［计］实现，履行，安装启用
epidemic [ˌepɪˈdemɪks]	*n.* 流行病，蔓延，时疫
proximity to	接近，临近
Civil Protection Act	民事保护法案
emergency planning	应急预案
infrastructure [ˈɪnfrəstrʌktʃə]	*n.* 基础设施，公共建设，下部构造
maintenance [ˈmeɪntənəns]	*n.* 维护，维持，保养，保持

Exercises

1. Explain the introduction to disasters and emergency management.

2. Disasters can be classified into natural hazards and technological or man-made hazards, what are the dissimilarities between them?

3. What is emergency planning? Explain the certain benefits for a municipality if there is an effective emergency operations plan.

4. What phases of the emergency management cycle are there in an emergency planning continual cycle?

5. Describe the four steps in the emergency planning process: the hazard analysis, the development of an EOP, testing the plan through a series of training exercises, and plan maintenance and revision.

科技英语论文的写作格式

科技论文是对研究成果的论述，按其目的可分为学位论文和学术论文。

1. 学位论文

在学位论文中，对科技英语的写作文体要求严格，分为前部、正文、后部三个部分。

前部 FRONT

封面	Front Cover
扉页	Title Page（Subtitle）
作者姓名/作者联系方式	Author's Name/Author's Address
分发范围	Distribution List
序言/前言	Preface or Foreword
致谢	Acknowledgments
摘要	Abstract
关键词	Keywords

目录	Table of Contents
图表目录	List of Illustrations (Tables, Graphs)
正文 MAIN TEXT	
引言	Introduction
分析/实验方法和过程	Analysis and/or Experiment Methods and Procedure
结果	Results
讨论	Discussion
结论	Conclusions (Summary)
建议	Recommendations
后部 BACK	
参考文献	List of References
附录	Appendixes
表	Tables
图	Graphs
索引	Index

2. 学术论文

学术论文是指发表在期刊上的论文和会议论文。

标题	Title
作者姓名	Author's Name
作者联系地址	Author's Address
摘要	Abstract
关键词	Keywords
引言	Introduction
分析实验方法和过程	Analysis and/or Experiment Methods and Procedure
结果	Results
讨论	Discussion
结论	Conclusions
致谢	Acknowledgments
参考文献	References
附录	Appendixes

Refuge Chambers in Underground Metalliferous Mines

Typically, irrespirable atmospheres result from fires in the workings, but they can arise from other causes, including outbursts of gases such as methane or hydrogen sulphide. The provision of refuge chambers is central to any emergency preparedness plan, which in turn is fundamental to the duty of care.

1. Nature of the Hazard

The hazards associated with an underground mine atmosphere becoming irrespirable due to ***contamination*** from fire or other sources are well recognised in the metalliferous mining industry in Western Australia. The training of miners has traditionally included various techniques of self-preservation using compressed air to create local breathable pockets in "blind" headings, vent tubing, safety helmets, etc. Such techniques date from the period when compressed air was the universal energy source in underground mining, and timber for sleepers, support or ***shaft*** furnishings was the principal combustible material.

The widespread use of diesel-powered and electrical equipment in underground mines means that compressed air reticulation systems have progressively disappeared, and the inventory of combustible materials has changed in both nature and quantity. Most mines now have significant stocks of diesel fuel, hydraulic oil, rubber (as tyres), polyvinylchloride (as cable sheathing and piping), and resin-based composite materials used for various machine enclosures.

Nearly all underground fires reported in Western Australia occur on vehicles, and may result from:

- high-temperature components on diesel prime movers providing ignition sources for oil sprays from leaking hoses;
- sparking from abraded Direct Current (DC) power leads damaging fuel lines;
- hot surfaces (i. e. > 350℃) such as exhausts and turbochargers;
- binding brakes causing grease fires in wheel hubs and igniting tyres.

The initial problem confronting an underground worker in the event of a fire is securing an immediate supply of breathable air. This is normally addressed by supplying everyone working underground with an oxygen-generating Self-Contained Self-Rescuer (SCSR). These devices come in various designs, and allow a person to travel from an endangered position to a safe haven. This presumes, of course, that such a safe haven exists in reasonable ***proximity*** to the endangered person.

The need for some form of refuge chamber in underground metalliferous mines has long been recognised. Early types were frequently a redundant excavation, which was blocked off to provide an enclosed space where the atmosphere could be overpressured using compressed air sourced from the mine system. This basic model has evolved to incorporate more functionality and increased sophistication. The increasing ***prominence*** of diesel-powered trackless equipment and a greater awareness of the needs of the workforce have provided the impetus to develop self-contained chambers that can be readily relocated to support the mining operation as it progresses.

There is uncertainty about whether or not the function of an underground refuge chamber can be extended to cope with a mine flooding or inrush situation—the two worst disasters in recent Western Australian mining history were caused by such events. There appear to be two aspects to this issue:

1) The design, construction and operation of such a chamber would ***be akin to*** that of a small submarine. The potential demands on a submersible chamber would require it to be purpose built because it would be impractical to adapt an existing fire refuge chamber.

2) Flooding or inrush events develop so rapidly that there is unlikely to be the opportunity for an underground workforce to move to a designated place of safety before being overwhelmed.

2. Location

(1) Distance from Workplace

Refuge chambers should be sited near active workplaces, taking into account the needs of people working there and potential hazards they face. It is recommended that the maximum distance separating a worker from a refuge chamber be based on how far a person, in a reasonable state of physical fitness, can travel at a moderate walking pace, using 50% of the nominal duration of the SCSR. If it is assumed that workers are equipped with SCSRs of nominal 30 min duration, at a rate of 30 litres per minute, then no one should be expected to travel further than 750 m to reach safety.

This distance should be regarded as an absolute maximum because:

- the duration of the SCSR can be adversely affected by the wearer's state of agitation;
- physical difficulties may be encountered while travelling;
- smoke from a fire underground may be so thick that crawling is the only feasible means of movement.

It should be noted that crawling is necessarily slower than moderate walking, and should be allowed for where applicable. Also, the ventilation practices at a mine may exacerbate the situation.

(2) Nominal Duration of SCSRs

The nominal duration of an SCSR is established at a specific rate of usage under standard conditions, as detailed in Australian Standard AS/NZS 1716: 1994. However, experience and experiments suggest that the rate of consumption is much greater under emergency conditions than might be expected. Arguments for more or better training, or both, and more frequent simulated emergencies have been advanced and have obvious value. However, the frequency of genuine emergencies involving the use of SCSRs is relatively low, and the financial impost of this training and simulation is significant.

3. Capacity

The primary function of an underground refuge chamber is to provide a safe haven for people working in the immediate area in the event of the atmosphere becoming irrespirable.

The chamber size should recognise that other personnel such as supervisors, surveyors, geologists and service technicians may also need to use the facility. The number of such people in the workings ***from time to time*** can require:

- provision for a refuge capacity more than double that determined from the size of the locally operating crew alone;
- implementation of a system to limit the number of personnel in the area.

4. Adapting Existing Facilities

The practice of designating a facility such as a lunchroom as a refuge chamber and equipping it for this purpose is a common, and traditional, response to the need to provide a safe haven. However, the size and general configuration of such a facility normally means that it can only be supported by the permanent services (ventilation, water and electricity) of the mine. In this scenario, therefore, these services must be immune from any interruption. From both technical and financial viewpoints, the equipping of such a resource with independent services is unlikely to be viable.

The lunchroom type of facility can be most useful when either performing as a fresh-air base or associated with one. In normal circumstances, a lunchroom is a semipermanent installation in a mine, and while it may be readily accessible from most areas, a maximum distance to all workplaces of 750 m is unlikely to be achievable.

5. Safety of Location

(1) Exposure to Hazards

A refuge chamber is perceived as the ultimate place of safety in an underground emergency. Its location should therefore be as secure from hazard as possible. Although the positioning of a refuge chamber is strongly governed by its accessibility for people in need of its protection, any potential susceptibility of its location to the hazards of rockfall, flooding, fire, explosion or damage from mine vehicles should be considered.

The placing of a refuge chamber close to installations such as transformer stations, explosives magazines, fuel storage facilities or vehicle parking bays should be avoided, as they are potential fire sources.

(2) Ground Conditions

While it is recognised that it may be impossible to locate a refuge chamber excavation in an area free from normal rockmass features such as faults, fractures and dykes, the susceptibility of these features to seismic activity or other ***disruptive*** influences should be thoroughly assessed. Major ground movements associated with seismicity can damage the chamber, its external service equipment, or restrict access to or from the chamber.

The ground support installed in the ***vicinity*** of a refuge chamber should be of a high standard, equivalent at least to the standard of permanent support as specified for the mine. Disused stockpile excavations, turning bays, redundant pump cuddies, and ventilation crosscuts have been variously used as sites for refuge chambers. The original purpose for which these excavations were made might have been designated as being temporary, and the ground support installed may reflect that status. Over time, rockmass conditions can deteriorate locally. Apart from posing a threat to the chamber and its associated equipment, poor ground conditions can introduce a hazard to personnel servicing the chamber on a routine basis, and people attempting to enter the facility for any other purpose.

(3) Water Make

A refuge chamber should not be placed in a location where water can accumulate in sufficient quantities to pose a risk to workers. Many chambers will be placed deep in the workings to be close to workers who might need them. Pump failure associated with an emergency can cause water to collect in the lower areas of a mine. Over a relatively long period of time, such as 36 hours, levels may rise sufficiently to reach deep refuge chamber positions. In this circumstance, it must be recognised that the existing water make of the mine can be seriously augmented by fluid from water mains damaged during an underground emergency.

New Words and Expressions

contamination [kənˌtæmɪˈneɪʃən] *n.* 污染，污秽，(语言的) 交感，混合

shaft [ʃɑːft] *n.* 竖井，井筒，(电梯的) 升降机井，通风井

proximity [prɒk'sɪmɪti] *n.* 邻近
prominence ['prɒmɪnəns] *n.* 突出，重要，卓越，出名
be akin to 与……同类的，近似的
refuge chamber 避难硐室
from time to time 时时，有时，偶尔
disruptive [dɪs'rʌptɪv] *adj.* 引起混乱的，扰乱性的，破坏性的
vicinity [vɪ'sɪnɪti] *n.* 附近，周围地区，邻近地区

Unit Eighteen

18

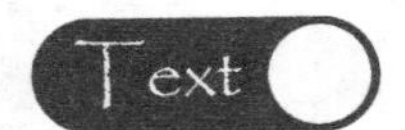

Food Safety

Assume that the 250 million people in the United States eat three meals a day. That means the nation's food safety policies directly affect Americans nearly 274 billion times a year—not including snacks. These policies have generally served us well. Food is relatively cheap, plentiful, and wholesome. In fact, the United States has by and large been thought to have the safest food supply in the world.

But perhaps it is not safe enough. Every year, enough ***contaminated*** food falls through the safety net to kill at least 9, 100 Americans and make at least 6.5 million others sick, according to researchers at the ***Center for Disease Control*** (***CDC***) . And that is only ***acute*** illness; the extent to which long-term disease is related to food is unknown. In addition, the social costs of food-borne illness, such as medical expenses and lost productivity, are sizable, estimated to reach between \$ 4 billion and \$ 8 billion annually.

The problem is not simply that individual food safety laws are not achieving what they were designed to achieve. Rather, most of these policies were created one by one to address specific problems, not in concert to achieve consistent, broad-based goals. Viewed in its entirety, the existing regulatory structure is inefficient, cumbersome, and costly. More important, it has not kept up with current needs and concerns. New scientific and medical knowledge, changes in trade and technology and consumer demographics and behavior have expanded the definition of "safe food" in ways that were never envisioned when the policies were created. In turn, the public has begun to raise legitimate questions about the government's ability to ensure the safest possible food supply.

Many Americans have come to realize that their outdated food safety system is not giving them their money's worth, and the last two decades have seen many calls for reform. But the government must do more than simply improve existing programs. Policy makers need to rethink the nation's overall approach to food safety regulation. Only when they define what role the government should play in food safety will they be able to determine what steps to take next.

1. A Century's Worth of Rules

Although growers, manufacturers, and retailers retain primary responsibility for the safety of their products, the federal government, in cooperation with state and local governments, ***keeps watch over*** the industry. A total of twelve federal agencies spend about $ 1 billion a year to ensure the safety and quality of the food we eat. Two organizations account for most of that spending: ① the Food and Drug Administration (FDA), which falls under the Department of Health and Human Services; ② the US Department of Agriculture (USDA), which includes five agencies that address food safety issues.

The federal government involves itself in virtually all stages of food production and marketing, from raw agricultural commodity to finished product. It sets standards for specific foods; approves food preparation equipment and processes; inspects facilities and products; sets legal limits for chemicals in food and tests food for compliance; regulates labeling and packaging; monitors state and local ***inspection*** programs; conducts research and consumer education efforts; takes action against illegal products; and monitors food-borne illnesses and other problems.

Obviously, this is a mammoth effort. Some 6,100 meat and poultry plants and more than 50,000 food establishments are subject to inspection by USDA or FDA. About 537,000 commercial restaurants, 172,000 institutional food programs, 190,000 retail food stores, and 1 million food vending locations submit to state and local inspection with FDA oversight. The government also keeps tabs on more than 70,000 separately labeled food products, 23,000 pesticides, 12,000 animal drugs, and thousands of additives—as well as $ 22-billion-worth of food and agricultural imports.

Its magnitude not with standing, this regulatory system did not develop under any rational plan. Programs emerged piecemeal, typically in response to particular health threats or economic crises. The earliest federal food safety laws, passed in the late 1800s, addressed such obvious problems as filth and fraud—for example, preventing manufacturers from adding impure or imitation ingredients to such products as tea and butter. Regulations were also designed to promote trade; for instance, meat and poultry inspection was introduced to certify the wholesomeness of meat exports. The first comprehensive federal food safety laws, *the Food and Drugs Act* of 1906 and *the Meat Inspection Act* of 1907, were intended to exclude misbranded or adulterated products from interstate commerce.

In the course of the 21st century, food production grew from a relatively simple, localized, farm-based industry into a multibillion-dollar enterprise. As food production and processing moved from the home to the factory, responsibility for food safety shifted away from consumers to processors, retailers, and in particular government regulators, whose role increased substantially. At the same time, scientists discovered that food can be contaminated not only by visible filth or impure fillers, but also by microorganisms (such as bacteria, viruses, and fungi), parasites (such as tapeworms); intentionally or unintentionally added chemicals (such as pesticides, animal drugs, flavor and color additives, industrial chemicals, or environmental contaminants); and natural poisons (such as the toxins in some fish). As understanding of food-borne hazards grew, so did concerns over food safety. Addressing one new worry after another, legislators amended old laws and enacted new ones. Today, a century's worth of such rules constitutes the complicated network that is our food safety system.

2. Inconsistency and Inefficiency

Food safety laws have unquestionably improved the safety and purity of the nation's food supply. But overall, the system suffers from its longstanding lack of coordination. The dozen federal agencies involved in food safety operate under different mandates and definitions. Too often, they duplicate efforts in some areas while ignoring others entirely. More important, their individual standards of risk are inconsistent.

The most obvious problems lie in the division of responsibilities between USDA and FDA. For the most part, USDA oversees products containing meat and poultry, while the FDA regulates all other food products. The arrangement is not quite as simple as it sounds. For example, the two organizations share jurisdiction for egg products. FDA also is responsible for products containing less than 3 percent raw meat or poultry as well as those containing less than 2 percent cooked meat or poultry. And both organizations monitor domestic and imported food for potentially harmful chemicals, such as pesticides, animal drugs, and environmental contaminants.

Yet these two organizations operate under substantially different statutory mandates. For instance, the USDA carries out a massive "continuous inspection" program at slaughterhouses, which by law may operate only when one of the department's 7,350 field inspectors is on duty. USDA also inspects all meat and poultry processing plants daily. By contrast, FDA inspects facilities under its jurisdiction, on average, once every three to five years. Due in part to budget constraints, FDA and state inspections cover less than one fourth of the nation's 50,000 food manufacturers, packers, processors, and warehouses each year.

The differences in the two organizations' approaches mean that food products that pose similar risks may receive widely varying scrutiny. For example, canned soup containing more than 2 percent meat poses essentially the same risk of contamination as canned meatless soup; in both cases, the health hazards rest not with the soup's ingredients, but with the canning process. Yet the USDA conducts daily inspections of the plant producing the soup with meat, while FDA may visit the plant producing the meatless soup only once every few years. Even without knowing what level of supervision is actually necessary, any observer can see that something is wrong: USDA is either wasting its time and money in daily inspections, or FDA is potentially allowing dangerous products to reach the market. Even as the inspectors concentrate on some products, they ignore other areas of equal or greater concern entirely. Fish—especially shellfish—caused 21 percent of all food poisoning cases arising from meat, fish, or poultry reported to CDC between 1978 and 1987. Yet seafood is not subject to mandatory federal inspection. In other words, the same system that requires continuous inspection of chicken practically ignores tuna.

The incongruities between USDA and FDA extend well beyond their inspection methods. For example, meat and poultry products must have a USDA stamp of approval for interstate sale, but food products under FDA jurisdiction generally require no pre-market ***certification***. USDA reviews construction plans for all manufacturing facilities for meat products, but non-meat food producers are not required to notify FDA about a plant's construction, or even its existence. And while USDA has legal authority to examine company records, FDA does not. As early as 1972, the Government Accounting Office (GAO) noted that this fact impaired FDA's ability to protect the public. Other GAO and congressional reports have suggested that FDA needs additional authority to halt the distribution of questionable products and to order recalls.

3. New Food-Borne Threats

Of the various sources of food contamination, microbes probably pose the greatest risk to human health. Harmful ***microbes*** in food cause nearly all cases of acute food-borne illness in the United States. Many cases, however, go undiagnosed, so that the actual figure is probably much higher than the conservative figure of 6.5 million annual cases—at least 24 million, according to an estimate by officials at FDA.

Most people have heard of salmonella. But scientists have lately identified other harmful organisms, such as listeria and campylobacter, as serious threats. This is due in part to better methods of detecting microbes, but it also reflects trends in food distribution that leave products vulnerable in new ways. For example, we depend on refrigeration to keep food safe in transport, but the listeria bacterium can survive ***refrigeration***. Each year, listeriosis strikes about 1,850 Americans; nearly one fourth of those people die. Similarly, campylobacter, the leading cause of bacterial diarrhea in the United States, tends to cause illness only when it reaches high levels in food. Improvements in packaging that allow longer food storage may enable the bacteria to grow to dangerous proportions.

4. The Original Food Safety Network Has Failed to Keep up with Changes

Even more worrisome is the appearance of new or stronger strains of contaminants. A generation ago, an uncracked egg was assumed to be a bacteria-free package; legislators responded by requiring cracked eggs—potentially infected with salmonella—to be used only in cooked products, because cooking destroys salmonella bacteria. Today, however, at least one strain of salmonella is able to pass from an infected chicken to a developing egg, so that eggs perfect to the eye might still be contaminated. Increased use of antibiotics in meat and poultry may also encourage the development of resistant strains of bacteria. While scientists believe microbes are today's chief food-borne threats, public attention tends to focus on pesticides, animal drugs, and other chemicals in food. Chemical residues may not make individuals fall ill immediately, but some people suspect them of causing cancer, birth defects, and other problems.

These types of contaminants can provoke outrage far out of proportion to the risks they pose. That is partly because many Americans view chemical contamination as an unnecessary risk, imposed on an unsuspecting population by food manufacturers who profit from the use of the chemicals. This perception surfaced in two episodes in 1989. First when consumer groups objected to the use of the pesticide Alar on apples, and later when import inspectors found some Chilean grapes tainted with cyanide. While no one became ill as a result of either one of these episodes, they caused great damage to consumer confidence and severe losses in the marketplace.

Yet USDA's methods employed in inspecting meat and poultry do not detect most microbial or chemical contamination. The standard inspection procedures of smelling, feeling, and looking at the product date from an earlier era when easily identifiable conditions, such as obvious disease or spoilage, were considered the chief dangers from these foods. Today, visible problems are minimal compared to the invisible threats, which can be detected only through laboratory analysis. USDA's grading standards for produce are equally out of date since they rely on criteria that are mostly cosmetic and may, if anything, encourage excessive use of pesticides.

Even if they had the resources, USDA and FDA could not identify all foods with illegal chemical residues and keep them from reaching the consumer. The government has no useful methods for detecting many of the residues it is supposed to monitor. Even where detection is possible, there are simply too many products to examine and too many contaminants to check for in the limited time before the product is sold and eaten. The government is seeking better ways of sampling and testing residues. But for now, government inspection may provide a false sense of security to those consumers who believe it means products are free of all contamination.

(1) Improved Technology

Technological advances in agriculture and the food processing industries have made it possible to offer a larger population a food supply that is cheaper, more varied, and more convenient than in the early 1900s. Yet some of the same tools that have dramatically expanded agricultural production—pesticides, fertilizers, and animal drugs—have themselves become cause for concern. Recent rules meant to ensure that newly introduced substances are safe to use have had the unintended effect of discouraging the development of safer chemical products; manufacturers and consumers instead stick with products approved under older, less stringent standards.

Mechanical improvements have introduced food safety problems, too. Traditional inspection methods cannot keep pace with high-speed equipment that allows only a few seconds for inspectors to examine each piece of meat and poultry. And the inability of inspectors to detect microbial contamination becomes even more worrisome in modern plants, where one infected chicken can swiftly contaminate hundreds of other birds processed with the same equipment. Better storage and transport means that food moves farther and faster than ever before—and, in turn, that a single source of contamination can affect more people in a larger area and in a shorter period of time. Given the sheer quantity of food in production, even small risks can cause harm on a huge scale.

In general, technology is raising new questions faster than regulators can answer them. For instance, some consumer advocates worry that new genetically engineered food products may cause unforeseen harm. While FDA is authorized to approve food additives, it has no comparable authority to review new foods before they enter the market. This issue drew attention in 1991, when Calgene, a California ***biotechnology*** firm, asked FDA to informally concur with its plans to market a tomato ***genetically*** engineered to remain firm during shipment. FDA is still reviewing the case.

(2) Changed Consumer Behavior

As the demographics of the population change, so does its risk of disease. Older people and people who are immune-compromised—two rapidly growing groups—are more vulnerable to food-borne illness than younger, healthier people. Eating patterns also shift with demographics. For example, Americans eat almost 60 percent more seafood now than they did ten years ago, partly because of growing numbers of minorities and senior citizens, who consume high proportions of fish. The risk has increased accordingly, as seafood is highly susceptible to contamination.

Changes in lifestyle make a difference, too. To meet consumer demand for low-processed, ready-to-eat foods, manufacturers are packaging more types of food than ever in convenient forms. Consumers, taking for granted that all packaged foods are safe, may overlook directions to refrigerate containers or to stir foods during microwave cooking—steps necessary to control microbes in certain products.

The trend toward eating out adds to the risk. USDA estimates that almost half the money consumers spend on food now goes to meals and snacks taken away from home. At the same time, budget constraints are limiting state and local inspection of retail food operations. Under-inspected establishments, such as popular self-serve counters at grocery stores, increase the risk of food contamination. Much of the responsibility rests with consumers themselves. As Americans have become further removed from the sources of their food, they have developed what may be a dangerous dependence on others to ensure the safety of their food. In general, Americans seem to have become increasingly unaware of the importance of cooking and storing food properly to destroy microbes and keep contamination from spreading. USDA estimates that 30 percent of all food-borne illnesses are due to unsafe handling of food products in the home.

(3) Broad-Based Reforms

The problems of the food safety network are far too broad and varied to be solved by narrowly targeted corrections. Real improvement will require large-scale reforms. A good way to begin would be structuring the network so it would work efficiently and confidently. Any change should begin with the two organizations that share most of the responsibility for food safety, USDA and FDA. The government has already taken some steps to ***clean up*** internal problems; in response to GAO's recommendations the secretary of agriculture announced in September 1991 that he would name a commission to consider how USDA can better manage cross-cutting issues within its own walls.

As for the FDA, the Edwards Committee report of 1991 recommended that the agency's status be elevated within HHS to put it on a par with corresponding regulatory agencies, such as EPA, FTC, and the Occupational Safety and Health Agency (OSHA). The committee also proposed that if HHS failed to act, Congress should consider restructuring FDA as a free-standing executive agency. HHS and the administration, however, received that proposal with little enthusiasm. Meanwhile, Congress is considering legislation to enhance FDA's enforcement authority.

New Words and Expressions

contaminated [kən'tæməneɪtɪd]	*adj.* 被污染的，受污染的，被感染的
Center for Disease Control (CDC)	疾病控制中心，防疫中心
acute [ə'kjuːt]	*adj.* 十分严重的，(疾病) 急性的，灵敏的，敏锐的
keep watch over	关注，留意，看守
inspection [ɪn'spekʃn]	*n.* 视察，检查
certification [ˌsɜːtɪfɪ'keɪʃən]	*n.* 证明，保证，检定
microbe ['maɪkrəʊb]	*n.* 微生物，细菌
refrigeration [rɪˌfrɪdʒə'reɪʃən]	*n.* 制冷，冷藏，[热] 冷却
biotechnology [ˌbaɪə (ʊ) tek'nɒlədʒi]	*n.* 生物工艺学，生物技术
genetically [dʒe'netɪkəli]	*adv.* 从遗传学角度，从基因方面
clean up	清理，大捞一笔

Exercises

1. What is the problem according to researchers at the Center for Disease Control (CDC)?
2. Which two organizations account for most of that spending?
3. Indicate the most obvious problems which lie in the division of responsibilities between USDA and FDA.
4. Describe the new food-borne threats.
5. Why has the original food safety network failed to keep up with changes?

科技英语中的摘要书写

论文的摘要是论文的重要组成部分，是论文的内容梗概。摘要以简明、确切的写作手法记述论文的主要研究内容，不加任何评论和注释性的词语。摘要的内容与论文的主要信息应相同，应使读者只通过对摘要的阅读，就可以对全文有一个大概的了解。

摘要主要由目的、方法、结果和结论四部分组成：

1）目的：简要说明研究的目的，说明提出问题的缘由，表明研究的范围及重要性。

2）方法：说明研究课题的基本设计，通过什么方法与途径，可以达到研究的目的。

3）结果：说明通过相应的方法与途径所得到的数据与结果。

4）结论：简要说明论文中所做工作有什么价值，是否可推广等。

1. 语法要点

1）在书写摘要中的研究目的时，通常使用一般现在时，或者一般过去时。例如：

The use of fluoride based foams increases the effectiveness of fire-fighting operations, but they are also accompanied by major drawbacks regarding environmental safety of perfluorinated compounds (PFCs).

As people's evacuation behavior in building fire is uncertain, there are therefore many uncertain factors affect the people's evacuation process.

To study the influence of cross wind on flame interaction and burning characteristics of two abreast liquid fuel fires, a series of burning experiments on transversely spaced heptane pools was performed.

2）在叙述所做的实验程序或实验方法时，应采用一般过去时。例如：

We conducted a retrospective review of 93 patients who underwent robotic-assisted colorectal surgery at our institution from 2012 to 2014.

3）叙述的内容是介绍数学模型、算法、技术时，采用一般现在时。例如：

In this study, Latin Hypercube Sampling method has been used to mathematically describe the determinacy and randomness in people's evacuation process.

4）在书写摘要中的结果时，应采用一般过去时态，当研究结果具有普遍推广性时，可采用一般现在时。例如：

Total heat and smoke release, and rate of smoke emission of the FPU/Alag composites were found to be greatly reduced in comparison with those of FUP.

The results of this study will have implications on safety distance designing and can help to increasingly understand the interaction process and mechanism of multiple pool fires from aspects of fluid flow and radiative heat feedback.

2. 常用句型

(1) 目的

A growing evidence base suggests that...

The purpose of this pilot study is...

In this study, ...

This paper proposes/treat/addresses/establishes/concerned with/considers/...
本文提出了/讨论了/论述了/论证了/介绍了/研究了
(2) 结果
The probability model developed in this study is...
This evaluation suggests that...
The results of this study will have implication on...
The paper concludes that...

Ensuring Biosafety through Monitoring of GMO in Food with Modern Analytical Techniques, a Case Study

For at least 10,000 years, crop cultivars have been modified from their original wild state by domestication, selection and controlled breeding, to become more pest resistant, to produce higher yields or to produce a better or different quality of product. In recent decades, the application of recombinant DNA technologies, including genetic cloning and transformation, has permitted the introduction of exogenous genes into unrelated species across species barriers. Thus, the term genetically modified has been introduced to describe an organism in which the genetic material has been altered in a way that does not occur under natural conditions of cross-breeding or natural recombination.

Sixty-seven million hectares of ***Genetically Modified*** (***GM***) plants were globally grown in 2003 and 99 % of global transgenic crop area was grown by six leading countries: USA, Argentina, Canada, Brazil, China and South Africa. The principal commercialised GM crops were soybean, maize, cotton and canola. The initial objective for developing GM plants was to improve crop protection; genetic modifications of GM plants performed up to now has mainly concerned the following traits: herbicide tolerance, Bt-derived (Bacillus thuringiensis) insect resistance, virus resistance, ***fungal*** resistance, male sterility/fertility restoration. The production of "next generation" transgenic crops is intended to generate products with enhanced nutritional value, durability, as well as those termed functional foods. The focus and emphasis is on applications with obvious benefits to consumers, an even more ambitious deliverable is the potential contribution to the solution of ***malnutrition*** in developing countries.

Though the benefits of this new technology have led to increased crop productivity, increased stability of production and some say a more sustainable agriculture and environment, a number of concerns have been expressed regarding its safety to both human health and the environment. The issues debated: allergenicity of GM foods to humans, horizontal gene transfer, outcrossing, loss of biodiversity, effects on non-target organisms, increased use of chemicals (Dale et al., 2002). The major concerns regarding foods risks are principally ***allergic*** reactions that new expressed proteins may provoke or the potential alterations of metabolism of GMO caused by genetic engineering.

Although cases of allergic effects (or reactions) have occurred, e. g. allergenicity caused by the expression of brazil nut transgene in soybeans, new regulations provide for the marketing of GM food only after a severe risk assessment has been passed. Another negative effect for human health may derive from the transfer of transgene or, of other parts of construct, to the cells of the body or to intestinal or food-associated bacteria (van den Eede et al., 2004). This would be particularly relevant if antibiotic resistance genes, used in creating GMOs, were to be transferred.

The risk associated with an intense agriculture of GM crops consists of the movement of genes from GM plants into conventional crops or related species in the wild which may cause loss of varieties. This risk is real as several countries have adopted strategies to reduce mixing, including a clear separation of the fields within which GM and conventional crops are grown. Another risk is represented by the possible effects on non-target organisms: For example, it has been shown that ***pollen*** of Bt-corn threatens Monarch butterfly larvae and aphids which have fed on Bt crops may affect ladybugs.

On the other hand, a certain concern has been expressed about the undesirable level of control of agriculture production by a few chemical companies; this impacts on farmers who become dependent from chemical industry for GM seeds supply as well as for chemicals in the case of herbicide-tolerant GM crops.

The global market value of GM crops, which is based on the sale price of transgenic seed plus any technology fees that apply, was estimated in 2003, to be \$ 4.50 to \$ 4.75 billion and is projected at \$ 5 billion or more, for 2005 (James, 2003).

By virtue of the ***precaution*** principle and of the possibility for consumers to make an informed choice between GM or not-GM food, a series of regulations have been introduced in Europe.

1. Methods for Detection of GMO in Food

The need to identify the specific GMO and to quantify the amount of GMOs in different food products demanded by the current EU legislation, has generated the requirement of sensitive and reliable detection methods applicable to both raw materials and processed products. Two scientific approaches for detecting genetic modification can be distinguished: DNA-based methods and protein-based methods. Currently available methods are based on the detection of specific DNA fragments by Polymerase Chain Reaction (PCR) or methods based on the detection of new expressed protein by Enzyme-Linked Immunosorbent Assay (ELISA). Besides PCR and ELISA, alternative methods, some based on already available techniques (western blot, southern blot, mass spectrometry, chromatography, near infrared spectroscopy, and surface plasmon resonance), and some novel ones (DNA microarrays) are being introduced. Here only PCR and ELISA methods will be dealt with as they are the most frequently used and have received much attention for validation.

(1) ELISA

Immunological techniques, currently used in diagnostic clinic, are now routinely used for the rapid detection of protein in novel food. Methods focusing on separation of proteins, such as Western Blot, are difficult for routine testing and time consuming. A variety of immunoas says formats have been developed to allow measurement of the primary binding reaction between antibody and its target antigen, the most common being sandwich ELISA, named in this way because the antigen (novel protein) binds both the immunosorbent (antibody bound to a solid support) and the labelled antibody. A sandwich ELISA with monoclonal antibodies against the protein CP4 EPSPS (5-enolpyruvylshikimate 3-phosphate synthase) in Roundup Ready (RR) soybean was tested and validated in a ring study coordinated by JRC of ISPRA, involving 38 laboratories (Lippet al. , 2000). The results were expressed in units of weight percentage for GMO in mixture with non-GMO of the same species. Assay data were given for a semi ***quantitative*** assay as a ratio between false positive and false negative samples, for a quantitative assay as an absolute error. Any sample containing RR soybean <2% was identified as below <2% with a confidence level of 99% , quantitative use of the assay measured on samples containing 2% RR soybean resulted in a repeatability and reproducibility of 7% and 10% , respectively, and detection limit of approximately 0.35% RR soybean. The method is applicable to samples where little or no treatment has been carried out during processing and thus the protein is not significantly degraded or denatured. The use of protein-based methods is limited to raw and partially processed materials. For a reliable quantification in complex matrices it is important to control the recovery of extraction and the matrix effect. Recovery of extraction can be quantified by spiking the samples under analysis with known quantities extracts of non-GMO containing materials at several levels across a quantitative range and measuring the recovery, thus evaluating interference, if any exists.

Furthers drawbacks linked to quantification are: expression below levels of quantification, expression only in specific part of the plant (tissue-specificity) or at different levels in correspondence of different plant parts or of different stages of plant development, limitation of quantification to only one-taxon. Nevertheless, advantages of ELISA methods consist in high degree of automation and high throughput of samples. Moreover, field variants, such as lateral flow strips kits, suitable for semi-quantitative test, are also available.

(2) PCR-Based Methods

High sensitivity and specificity of PCR-based methods make them, at present, the most common tool in GMO detection (Holst-Jensen et al. , 2003). A prerequisite for GMO detection by PCR is at least the partial knowledge of the target gene sequence in order to design the two primers that will function as the trigger for the polymerisation reaction leading to the amplification of a DNA fragment between the primers. PCR primers can be directed to different sequences of the introduced DNA. Generally, the gene construct is composed of several elements: ① genes of interest; ② regulatory sequences, most used are the 35S promoter (P-35S) of Cauliflower Mosaic Virus and the Agrobacterium tumefaciens nopaline synthase terminator (T-Nos); ③ marker genes, e. g. genes conferring resistance to antibiotics like neomycine 3'-phosphotransferase, conferring resistance to kanamycin (marker will be avoided by legislation from 31 December 2004 according to Directive 2001/18/EC); ④ DNA from the cloning vector. At least four categories of GMO tests, corresponding to different level of specificity, can be recognized (Holst-Jensen et al. , 2003): screening, gene-specific, construct-specific and event specific methods. Screening methods target the most common sequences present in GMOs currently on the market, e. g. P-35S, T-Nos, nptII, gene specific methods target the gene of interest, construct-specific methods target junctions between adjacent elements of the construct, for example a region spanning the promoter and the gene of interest. Event-specific methods are "line specific" (transformation event specific) because the target is the junction between the host genome and the inserted DNA.

In conventional PCR, after ***amplification***, PCR products are usually subjected to agarose gel electrophoresis to separate fragments according to size and then visualized by staining with an intercalating dye, e. g. ethidium bromide. Different methods can then be used to confirm the PCR results: restriction analysis, sequencing, hybridisation, and nested PCR.

Compliance with the 0.9% mandatory labelling, makes GMO quantification a crucial step in GMO analysis. This can not be addressed with qualitative PCR because the amount of PCR product generated by amplification is poor indicative of the initial target concentration due to the "plateau phase" reached by conventional end-point measurement. At present, to overcome this problem, two PCR techniques are being used: quantitative-competitive PCR (QC-PCR) and, most commonly, real time PCR (RT-PCR).

QC-PCR is based on the co-amplification of the target DNA with an internal standard which, has the same sites of primer binding respect to the target, but is distinguishable from the target by size difference. By running a series of experiments in which each sample is co-amplified with varying amounts of the competitor, it is possible to determine the initial amount of template DNA. As long as amplification efficiencies of the target and the competitor are the same, if the resulting PCR products are of equal intensity, the initial concentration of target and internal standard are equal (equivalence point).

This technique has the advantage to be applied in laboratories that already use qualitative PCR without the need of specialised equipment, however drawbacks are: difficult to obtain standardization, time-consuming, significant risk of carryover contamination due to extensive handling and pipetting.

In the RT-PCR system, the increment of PCR product is measured by monitoring the increase of fluorescent signal emitted by specific fluorescent probes present in the amplification reaction. Fluorescent signals emitted from irradiated samples are detected by a CCD (Charged Coupled Devise) camera and converted into quantitative estimates by software.

In contrast to the end-point measurements, RT-PCR allows for the monitoring of the amplification reaction as it actually occurs (in real time), and analyzing data when the reaction is in its exponential phase (log phase). This phase is used because the amplification efficiency remains constant, and there is a direct correlation between the amount of PCR product generated and the initial amount of target template. It is therefore possible to quantify the DNA content of a sample by simple interpolation of its threshold cycle (Ct), defined as the number of cycles necessary to generate a signal statistically significant above the noise level, with a log standard curve plotting the initial amount of target molecules versus the Ct value. Several different fluorescent systems are available: intercalating dyes (e. g. SYBR Green) and hybridisation probes: hydrolysis probes Fluorescent Resonance Energy Transfer (FRET), molecular beacon and scorpions probes.

Advantages of this methodology are the following: the applicability also in high processed foodstuffs due to amplification of very short DNA fragments, high specificity thank to the use of target specific probes, reduced time for analysis due to the ***elimination*** of gel electrophoresis step, possibility of automation. On the other hand, the major drawback is the high cost of instrumentation and reagents.

2. Method of Analysis

The application of GMO analytical methods for official control according to the present European regulation requires the use of validated procedures. Validation is the process of proving that an analytical method is acceptable for its intended purpose. In general, methods for must include studies on: applicability, practicability, specificity, robustness, accuracy, sensitivity, repeatability, detection limit (LOQ) and quantification limit (LOQ). Standardization within Europe is being undertaken by bodies such as the European Committee for Standardization (CEN), which deals with the standardization of methods, and the International Standardization Organisation (ISO), dealing with harmonisation of standards. Currently, a database of validated methods compiled by JRC is available.

New Words and Expressions

genetically modified (GM) 基因改造，转基因的
fungal [ˈfʌŋg(ə)l] *adj.* 真菌的，真菌引起的
malnutrition [mælnjʊˈtrɪʃ(ə)n] *n.* 营养不良
allergic [əˈlɜːdʒɪk] *adj.* 过敏的，变态反应性的，变应性的，过敏性的
pollen [ˈpɒlən] *n.* 花粉
precaution [prɪˈkɔːʃ(ə)n] *n.* 预防，预防措施，防备，避孕措施
quantitative [ˈkwɒntɪˌtətɪv] *adj.* 数量的，量化的，定量的
amplification [ˌæmplɪfɪˈkeɪʃən] *n.* 扩大，扩充，(声明等的) 补充材料
compliance with 符合，遵守
in contrast to 与……相反的
elimination [ɪˌlɪmɪˈneɪʃən] *n.* 消除，消去，弃置

Unit Nineteen

19

Road Safety

Although the road system offers a high degree of mobility in Asian cities, a large amount of personal trauma is experienced through crashes. The term "crash" is preferred to "accident" since accident implies accidental and these events are unavoidable and non-preventable. Crashes are considered a failure of the system that can be eliminated through good engineering.

1. The Traffic System

The traffic system comprises three major components: humans, vehicles, and roads (see Figure 19-1). Both human (drivers or controllers) and vehicle (moving parts) characteristics are essential elements for road design and management. The properties of both elements should heavily influence the design and safety of roads.

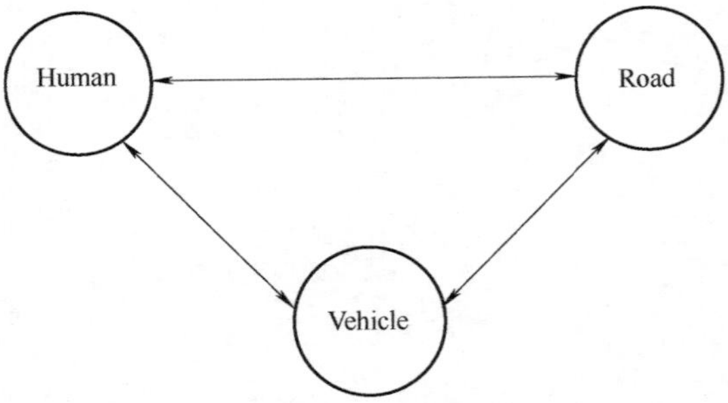

The road element is the primary concern of the traffic engineer, but ***familiarity*** with the other components is essential. All three elements must be fully compatible for the system to operate without failure.

Human

Road

Vehicle

Figure 19-1 The Traffic System

(1) Human

Humans control the vehicle, taken as given (static), but there must be detailed knowledge of the characteristics (e. g. , vision, memory, reaction time, judgment, ***expectancy***, information processing, and sensing), which may differ by age, gender, and experience. These lead to rules (e. g. , drivers should be confronted only with one simple decision at one time). Human factors should influence design geometry, ***alignment***, signs, signals, etc. Personal characteristics like eye height, vision range, etc. , need to be identified. Generally, we design for the 85th percentile, to ensure both safety and economic performance.

(2) Vehicle

Vehicles operate in road environments and their characteristics need to be considered to ensure that they can safely operate. Vehicle dimensions such as length, width, and weight need to be identified and considered for safe road design and management. Performance of vehicles, such as acceleration, cornering, and braking, also needs to be considered.

(3) Road

The road environment is the main concern of transport engineers. Road design and operational management are important areas that have a large influence on road safety (Underwood, 2006; Ruediger and Mailaender, 1999; Ogden, 1996). Although driver and vehicle factors are often identified as the cause of crashes, a significant proportion of road crashes is attributable to the road environment.

Road crashes occur when a number of adverse factors combine to cause a failure of the traffic system. Although the dominant factor in road crash causation is often the driver, the road environment often significantly contributes to ***likelihood*** and severity of road crashes.

Although there have recently been a number of initiatives to improve the performance of vehicles as well as increase the awareness of road safety and responsibility of drivers, techniques aimed at improving road conditions will be emphasized here.

It is important to determine an appropriate functional classification for roads in a network. Roads have two primary road functions: through movement and property access. However, these are conflicting! Safety problems typically occur when roads attempt to serve both functions. Contemporary road network planning emphasizes the separation of functions (see Figure 19-2).

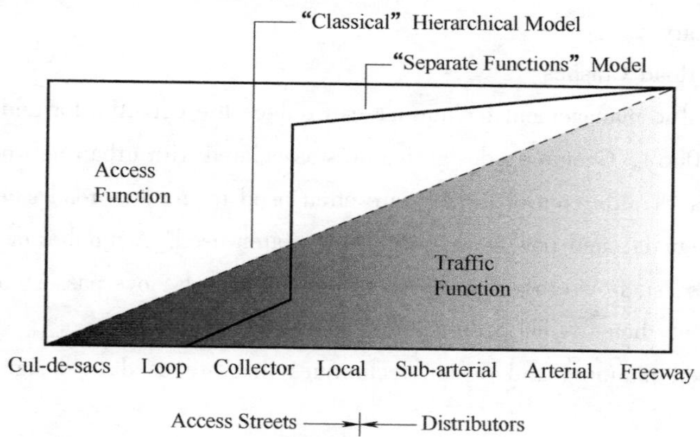

Figure 19-2 Functional Mix of Roads and Streets (Source: Ogden, K. W., and Bennett, D. W., 1989, *Traffic Engineering Practice,* 4th ed. Clayton, Australia: Department of Civil Engineering, Monash University, Chapter 6.7)

2. Hazardous Road Locations

(1) Identification of Hazardous Road Locations

Several statistics are generally used to identify elements of the road network that have a high crash frequency or severity levels.

Black spots are ***clusters of*** road crashes occurring at precise locations identified by features of road geometry such as junctions and bends. Black sites (specific lengths of road) and black areas are other locations of high risk, which may be identified individually from accident history in terms of clusters of accidents.

The concept of exposure is important when comparing crash sites. Exposure is a measure of the opportunity for a crash to occur (e. g., traffic volume or VKT). Network elements are often analyzed by

- crash type—classification of vehicle movement or road user involved;
- crash rate—number of crashes per unit of exposure;
- crash severity—a measure of the consequence of the crash in terms of the casualty sustained;
- casualty class—a measure of the consequence of a crash in terms of the number of injuries.

(2) Diagnosis of Road Crash Problems

Collision diagrams are generally used to investigate the cause of crashes as part of the in-office analysis. This summarizes all vehicle movements involved in collisions and highlights ***predominant*** crash types and vehicle ***maneuvers***. Other details from the crash report, such as the date and time of the crash, condition of the road, light conditions at time of the crash, geometry of the site, and type of vehicle involved, are usually analyzed.

An on-site inspection is necessary to assess road conditions and other site factors accurately. This should identify adverse features of road design and adverse environmental features.

Other factors influencing behavior at the site, such as the physical and mental condition, experience, and age of drivers should also be investigated. When selecting countermeasures, a large degree of judgment and experience is necessary.

(3) Prevention of Road Crashes

Road design and traffic management techniques can reduce the potential for and the occurrence of road crashes (Underwood, 2006). Geometric design features associated with urban networks such as intersection design can cause crashes. Traffic control devices are often used to improve road safety. Signs (regulatory, warning, and guide), signals, and pavement markings are often used. A number of treatments can increase protection for pedestrians, such as crossings, medians, refuge islands, overpasses, and underpasses. Street lighting also provides safety benefits for pedestrians.

There are specialized planning and design techniques that are used to design safe street systems in residential areas.

Governments need to develop a works program to allocate funds to improve road safety. This requires individual projects to be ranked due to budgetary constraints. The aim of the road safety program is to maximize economic benefits. It is common to use standard economic evaluation criteria such as the Benefit-Cost Ratio (BCR) or Net Present Value (NPV) to rank projects.

Implementation cost estimates are based on typical treatments and experience, including maintenance costs. Estimates need to be made for each countermeasure and site. Here, a crash reduction factor (the expected reduction by countermeasure by crash type) needs to be estimated and is generally based on previous performance. Road crash costs are generally ***tabulated*** as per person by severity or per vehicle. These need to be converted to per-crash cost.

A process for determining whether or how effectively a road safety activity has brought about the desired result is necessary for ensuring financial accountability as well as establishing causal relationships. This process relies heavily on crash data and statistical testing.

3. Outlines of Traffic Safety Issues

(1) Trends and Causes of Accidents

It is clear that the lack of improvement in traffic safety is one of the biggest issues from which we are suffering all over the world. Due to the recent intensive progress in economies and industries in most Asian countries, rapid ***motorization*** can be observed and thus the number of traffic accidents tends to increase year by year. Accordingly, it is very important for us to propose effective countermeasures to mitigate traffic accidents and to do the related basic research, especially on both mechanism and impact of traffic accidents.

Although it seems that Japan is facing a relatively mature era in terms of motorization compared with the other Asian countries, there is still an increasing trend of the number of accidents and persons injured as shown in Figure 19-3.

Number of:	1995	2006
Fatalities	10,679.00	6,352.00
Injured persons	922,677.00	1,098,199.00
Accidents recorded	761,789.00	886,864.00
Vehicle registered	70,100,000.00	79,452,557.00
License holder	68,563,830.00	78,798,821.00

Figure 19-3 Statistics Related to Traffic Accidents in Japan (1995 versus. 2006)

Figure 19-4 shows the major causes of traffic accidents reported by the national police agency in Japan. As shown in the figure, most accidents might have been caused by human error, including late awareness of dangerous events in traffic and mistakes in both judgment and maneuvers. Accordingly, it can be said that it is necessary for us to enhance the basic research on the causes and mechanism of accidents in order to make countermeasures for traffic safety more effective.

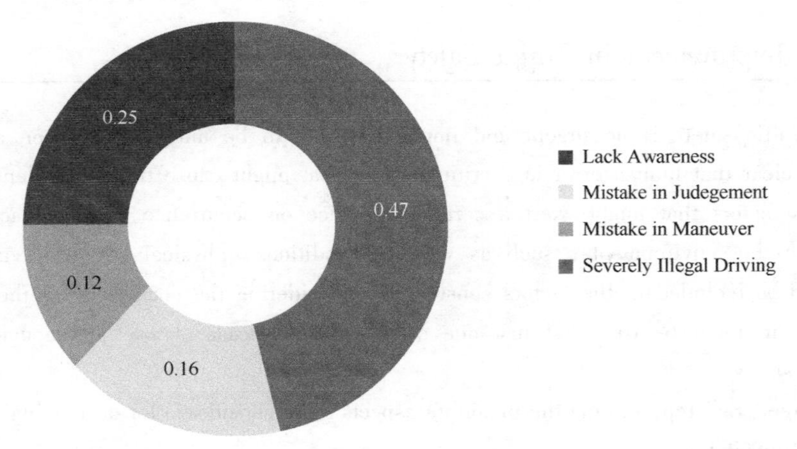

Figure 19-4 Major Causes of Accident in Japan

In addition, we have to ***pay attention to*** the high percentage of elderly persons (above sixty-five years) and young generations (below twenty-four years) among fatalities in Japan.

Considering the serious situation in terms of traffic safety in Japan, the prime minister provided us with the following statement in January 2003: "We are aimed at achieving the safest roadways all over the world and making the number of fatalities less than 5,000 per year during the next ten years." Based on this statement, the Japanese government has initiated various countermeasures to improve traffic safety, such as improvement in traffic regulation and strict enforcement, improvement in traffic safety devices, enhancement of passive safety of vehicles, and so on.

(2) Difficulty of Traffic Safety Studies

The ultimate goal of traffic and transportation engineering is to make traffic flow on road networks more efficient, smoother, and safer. Compared with traffic safety aspects, much research has been carried out for evaluating the level of service provided by a road transportation system from the viewpoints of efficiency and smoothness of traffic flow, and several indices are available—for example, travel speed, jam length, maximum throughput of a network, and so on. In past studies, safety has been frequently discussed on the basis of the change in the number of accidents recorded. However, it seems to be difficult to evaluate the safety aspect of traffic flow based only on the data of recorded traffic accidents in an objective or statistical manner, because accidents are a rare event when we focus on a certain road section.

As mentioned in the previous subsection, it can be said that human error might be a predominant cause of accidents. For example, in Japan, around 75% of accidents are caused mainly by human error. Accordingly, it is expected that the enhancement in basic research on both causes and mechanisms of traffic accidents might contribute to improvement in countermeasures for traffic safety. However, there have been difficulties in collecting suitable data for analysis of mechanisms of traffic accidents. This might be a reason why less accumulation of safety research can be found compared with that of traffic flow analysis. Recently, as mentioned in the next subsection, rapid progress in ***Intelligent Transportation Systems*** (***ITS***) has led to a drastic change in data collection, and it has become possible for us to collect data directly on the occurrence of accidents and to collect data on interaction between vehicles and corresponding conflicts. The innovative changes in data collection will bring about better understanding of the mechanisms of traffic accidents through accumulating the basic research.

4. Its for Improvement in Traffic Safety

Improving traffic safety is an urgent and important issue to be mitigated as soon as possible. As mentioned, it is clear that human error is a critical factor that might cause traffic accidents. In addition, there are various factors that might exert a certain influence on occurrence of traffic accidents. Traffic conditions and driving environments, such as weather conditions, physical structure, and geometry of roadways, should be included in the factors considered. Considering the complexity in the mechanisms of traffic accidents, it might be true that just one or two countermeasures are not enough for drastically improving traffic safety.

Here are three "Es" representing the important aspects to be considered for developing countermeasures for improving traffic safety:

- education;
- enforcement;
- engineering.

Road transportation is one of the important social systems in enhancing socioeconomic activities. In other words, everybody belonging to our modern society is required to participate in the road transportation system as its users. In this sense, enhancement in education related to traffic safety is very important in terms of providing people with both basic knowledge about traffic safety issues and opportunity to review their behavior from the viewpoints of traffic safety in the long run. But actually there seems to be a lot of room for improvement in terms of traffic safety education at the elementary or ***junior*** high school level in Japan. No systematic education has been provided to students considering the educational guideline issued by the Ministry. This is one of major issues to be discussed intensively.

Enforcement is expected to play an important role in directly reducing the effects of accidents. The following are examples of recent changes in traffic regulation aimed at improving traffic safety in Japan:

1) 1996: Compulsory usage of child safety seat.

2) 2002: Strict enforcement of regulations for drunk driving with severer fines and penalties.

3) 2004: Strict enforcement of regulation for mobile phone usage.

Especially, it seems that the change in regulations in drunk driving and their strict enforcement might have contributed to reduction in the number of accidents caused mainly by drunk driving.

New Words and Expressions

familiarity [fəmɪlɪ'ærɪti] n. 熟悉，通晓，亲近，认识

expectancy [ɪk'spekt (ə) nsi] n. 预期，期待，预料，盼望

alignment [ə'laɪnm (ə) nt] n. 排成直线，(国家、团体间的) 结盟

likelihood ['laɪklɪhʊd] n. 可能性

cluster of 很多，一组，一串

predominant [prɪ'dɒmɪnənt] adj. 显著的，明显的，盛行的，占优势的

maneuver [mə'nʊvə] n. 演习，花招，谨慎或熟练的动作

tabulate ['tæbjʊleɪt] v. 制……的一览表，使成平面

motorization [ˌməʊtəraɪ'zeɪʃən] n. 动力化，电动化

pay attention to 注意，重视

Intelligent Transportation Systems (ITS) 智能交通系统，智能运输系统

junior ['dʒuːnɪə] adj. 地位 (或职位、级别) 低下的，青少年的

Exercises

1. Explain the characteristics of traffic system.
2. What indicators are analyzed for network elements to identify hazardous of Road Locations?
3. Indicate the most obvious problems which lie in the division of responsibilities between USDA and FDA.
4. Describe the outlines of traffic safety issues.
5. Find some traffic cases and give your opinions on how to improve traffic system.

科技英语的结论书写作

科技论文的结论放在正文的最后，是对整篇文章的总结，是对实验、理论分析和观测所得到

的结果经过判断、推理、归纳等，在整篇科技论文中占有很重要的地位。要想写好结论，就要分清摘要、引言和结论之间的关系，要搞清楚各自的特点。

撰写结论时要注意以下几个问题：

1）用词要准确。在结论中尽量避免出现“也许”“大概”“可能”这些模棱两可的词汇。

2）不能简单重复前文的小结。结论不是对文中各章节的结论进行简单复制，而是对全文整体内容的总结，表明全文的中心思想，也不要书写前文不曾得到的结论。

3）名词要解释清楚。比如前文有介绍过一些名词的简写字母，在结论处不能以简写的形式出现，而要说明全称。例如：不能用“FTA”，而要以“事故树分析法”代替。

4）不要出现评判性的语言。结论中不要批评和否定别人的成果，也不要进行自我批评和自我表扬，只要准确说明自己的研究成果即可。

5）不要出现“通过上述分析，得出如下结论”这样的语句。

1. 语法要点

1）当结论具有普遍性时，采用一般现在时。

A wide range of applications of destress blasting is presented in this paper and it can be summarized that destress blasting, if properly applied, can help mitigate the problems of high stress and associated coal bumps and rock bursts.

The shock wave is reflected by the chamber and the reflected wave pressure on the impact surface is about two times higher than that on the incident wave pressure.

2）当结论只适合本文所涉及的特定条件，只对于本文有效时，采用一般过去时。

The analysis results show that unsafe behaviors existed in and had precipitated influence to all gas explosion accidents, and clearly analyze the specific unsafe behavior and improve it to prevent gas explosion effectively.

The methodology for mining environment estimation proposed in this study was implemented in the hybrid system PROTECTOR, a helpful tool for assessment of the general mine safety state and category of hazard in mining environment.

3）在叙述进一步的研究方向或本文结论的应用前景或效果时，采用一般现在时，并且加入情态动词，如“may”“should”“could”等。

It's application in open-pit mine will help to manage unsafe factors, reduce danger, establish long-term safety mechanism and eventually improve the level of intrinsic safety of open-pit mining system.

2. 结论常用句型

In this paper, ...

The following important points should be considered for...

The research of... is a new field of...

The results indicate/prove/show/reveal that...

It should be noted that this study has examined only...

The study has addressed only the question of...

The experiments mentioned in the paper confirmed at...

We have demonstrated in this paper...

It is shown from the numerical simulation that...

An Examination of the National Road-Safety Programs in the Ten World's Leading Countries in Road Safety

Road accidents incur significant human, social and financial costs. According to the World Health Organization, every day, more than 3,000 people in the world die in road accidents, while the global financial cost of road accidents is estimated at 518 billion dollars a year. Moreover, according to global estimations, by the year 2020, road accidents will reach third place among ten leading death causes in the world. ***In light of*** this tragic data, many countries around the world search for measures and interventions aiming to reduce the road accidents' burden.

Over the last decade, a remarkable decrease in the number of road traffic fatalities was observed in many advanced countries, in which usually national road safety programs were applied. According to international experience, development and implementation of a national road safety program constitutes a key component in the country safety management policy that contributes to raising awareness to the road safety field, building partnerships to ***tackle*** the problems and focusing the intervention efforts on the issues required. Today, both European and world-wide organizations encourage countries to develop and implement national road safety programs as a background condition for systematic work on promoting road safety. In a more general context, currently the international research seeks to define essential components of effective road safety management, where national road safety program is well-recognized as one of them.

This study intended to examine the National Road Safety Programs (NRSP) that were carried out in the ten world's leading countries in road safety, in order to identify the most effective interventions that contributed to improved safety in those countries and to examine the possibilities of their application in Israel. The topic of NRSP is not new for Israel where one of the first NRSP was developed in the 1990s; and the latest NRSP was adopted by the government in the mid 2000s. The study was initiated by the National Road Safety Authority, based on the reasonable ***assumption*** that learning from the experience of other countries on treating road accidents can contribute to road safety in Israel by means of including these interventions in the development of new safety programs in Israel. Such an assumption is common in road safety work, where many reports summarize good-practice experiences on various safety-related issues. However, the difference of the current study was in a more purposeful definition of its framework: A detailed reviewing of national road safety programs of the best-performing countries aimed at creating a comprehensive list of safety interventions that were adopted by a significant part of those countries and, according to the available evidence, probably contributed to their road safety success. It was believed that adopting similar interventions by a future Israeli NRSP may improve the program's potential in reducing road accident injury and save resources needed in the course of the program's development.

The study was carried out mostly in 2009 (the final report was published in the middle of 2010) and, therefore, based on data and publications accessible then. However, the main value of the study can be seen in the rational of the analysis performed and its general findings which might be of interest for other countries seeking to develop a NRSP.

1. Selecting the Ten Leading Countries in Road Safety

In this study, the conventional way of comparing safety levels of countries was applied, by using general safety indicators: number of road fatalities per population, vehicle fleet and kilometers-traveled. The three indicators allow pointing out the "safest" countries in terms of the risk of being killed in traffic accidents, related to various forms of exposure, and are recommended to be applied together in order to overcome the weaknesses of each of them.

An additional way of comparing country progress, which was applied in Europe over the last decade, is to examine over-time changes of a certain figure, e. g. the absolute numbers of road traffic fatalities in a country. For example, the countries achieving the European Union (EU) goal of reducing the annual number of fatalities by 50% between the years 2000 to 2010, were France, Portugal and Luxembourg, which were characterized by an 8%–10% average annual reduction of fatalities in years 2001–2008.

Thus, to select ten leading countries in road safety, European and other developed countries were compared in terms of four indicators: ① road fatalities per population, in 2008; ② the average annual percentage change in road deaths, in 2001–2008; ③ road fatalities per million passenger cars, in 2006; ④ road fatalities per 10 billion passenger-kilometers traveled, in 2006. The data for these figures were collected from publications of international organizations, while data availability dictated a comparison year for each indicator considered. Following the study objective, the leading countries were selected compared to Israel's position according to each one of the indicators considered.

To summarize both examinations, a country definitely selected in a certain case received the score "1", where one partially fitting received the score "0. 5". Subsequently, two weightings were performed given equal and unequal weights to each examination. Both weightings produced similar results, recognizing eight leading countries with highest final scores: the Netherlands, Germany, Switzerland, France, the United Kingdom, Sweden, Norway and Belgium (Luxembourg was omitted from the list as a country not having a NRSP). Further, from the pair of Finland and Denmark with equal scores, Finland was selected as having better safety indicators than those of Denmark. Similarly, from the pair of Spain and Portugal with equal scores, Portugal was selected due to a significantly higher percentage of annual fatality reduction. Hence, the ten leading countries in road safety performance selected for further examination were: the Netherlands (NL), Germany (DE), Switzerland (CH), France (FR), the United Kingdom (UK), Sweden (SE), Norway (NO), Belgium (BE), Finland (FI), and Portugal (PT).

2. Detailed Examination of the NRSPs of Leading Countries

The characterization and analysis of the NRSPs in the selected countries were based on the national programs' documents published in these countries, together with country's self-reporting to international bodies, follow-up reports and other papers published during the programs' implementation. The information sources, for each country, were found through literature surveys, screening publications of international bodies and direct contacts with researchers working in the countries.

(1) Typical Safety Problems

As expected, a considerable similarity was found between typical safety problems in most countries considered. Thus, to summarize the findings from different NRSPs, a uniform classification was applied. The safety problems were divided into eight areas: vulnerable populations, drivers, behaviors, ***infrastructure***, vehicle conditions, information and knowledge, safety management, and rescue services, where each area included a further division into categories (specific safety problems).

It demonstrates the over-time changes in importance of selected safety problems, where among the most common safety problems in the ten countries appear:

1) In the area of behaviors: speeding, driving under the influence of alcohol, nonuse of vehicle restraints, failure to ***comply with*** traffic laws, driving under the influence of drugs and medications, both over the past decade and in new programs.

2) In the area of vulnerable populations: children and cyclists, over the past decade, when recently the focus has moved to motorcyclists, still together with children.

3) In the area of drivers: young novice drivers, drivers with repeated offenses and all drivers were in the focus over the past decade, while recently the focus remained mostly on young novice drivers and drivers with repeated offenses.

4) In the area of infrastructure: safety of urban streets, single-carriageway roads, design guidelines, roadside obstacles, over the past decade, while recently the focus remained mostly on single-carriageway roads and urban streets.

5) In the area of vehicles: vehicle maintenance problems were common over the past decade, where today the emphasis is mainly on non-implementation of vehicle safety measures.

6) In the remaining areas, no common problems were found for most countries, although, both in the past and today, four countries underlined the need for reducing arrival time of rescue services.

(2) Safety Interventions Implemented to Address the Problems

The classification introduced also served as a basis for summing up safety measures/interventions recommended for implementation by the countries' NRSPs. For each safety problem defined, a list of treatments was composed, both for previous ("last decade") and current safety programs, where ***safety interventions*** are ranked according to their appearance frequency and, for each item, relevant countries are mentioned.

Based on the summaries produced, it was seen that the most common interventions implemented by the NRSPs in the past decade were:

For "Speeding": camera-based speed enforcement (appeared in 10 countries); publicity campaigns emphasizing the relationship between speed and safety (in 5 countries); fine increase (in 4 countries); legislation changes supporting automatic enforcement (in 3 countries).

For "Nonuse of vehicle restraints": enforcement of safety belts' use and ***helmet*** wearing (in 7 countries) and publicity campaigns (in 6 countries).

For "Urban streets": creating 30 km/h zones (in 6 countries); traffic calming measures (in 3 countries in the past decade and also in 3 updated programs), where other measures also include elements of traffic calming and/or separating vulnerable road users from vehicle traffic.

For "Driving under the influence of alcohol": increased enforcement-performing breath tests on roadsides (in 6 countries); publicity campaigns (in 5 countries); stricter punishment (in 5 countries) and lowering the maximum legal blood alcohol concentration (BAC), from 0.8 g/l to 0.5 g/l (in 3 countries).

For "Rural single-carriageway roads": building a physical separation between travel directions, i.e.

"2 + 2", "2 + 1" roads (in 4 countries) and other infrastructure improvements (in 3 countries).

For treating "Roadside obstacles": installing guardrails and removing roadside obstacles (in 5 countries).

For "Children" as a vulnerable population group: safety education at schools (in 3 countries).

For "Young novice drivers": graduated driving license (in 5 countries) and stricter punishment for young driver offenses (in 4 countries).

Regarding the "arrival time of rescue services", promoting eCall installations in vehicles appeared in two NRSPs in the past decade and in three new ones. However, for one of the most common problems among the countries-motorcyclists a prominent solution was not found, while various measures were applied or suggested by different countries. Among other safety problems specified in the program's updates for which prominent interventions are lacking can be mentioned: the problem of "driving under the influence of drugs/medications", "driving fatigue", and the elderly population.

New Words and Expressions

in light of	根据，鉴于，从……观点
tackle [ˈtæk (ə) l]	*v.* 处理，阻截，与某人交涉，向某人提起
assumption [əˈsʌm (p) ʃ (ə) n]	*n.* 假设，假定，担任，(责任的) 承担
infrastructure [ˈɪnfrəstrʌktʃə]	*n.* (国家或机构的) 基础设施
comply with	服从，遵从，应，顺应，照办
safety intervention	安全干预措施
helmet [ˈhelmɪt]	*n.* 头盔，防护帽

20

Unit Twenty

Evacuation from Trains—The Railway Safety Challenge

1. Evacuation—What's the Problem?

Severe accidents in transport systems such as railways means mass ***evacuations*** often under time pressure, with immediate threats and in difficult circumstances, e. g. in case of a fire or if the evacuation must take place in a tunnel or on a bridge. The frequency of such events is usually low but the consequences can be severe. However, mass evacuations occur quite frequently in situations where a single or several trains are stopped because of track, vehicle or traffic management problems. Even in these situations passengers and staff are exposed to potentially negative consequences, such as proximity to electrical lines or the risk of being hit by a passing train if the passengers enter a non-secured railway track.

2. Research on Evacuation from Trains

Research on evacuations from buildings has been more extensive as compared to research in transportation and has thus been used as a basis for developing scientific knowledge on evacuation behaviour. Mass evacuations in situations with severe threats have been studied in railway accident investigations and also in evacuation exercises in difficult circumstances such as in subways and road tunnels. Until now, there has been no systematic gathering of data about how passengers and different professional companies handle evacuation situations from trains. One of the aims of the project has been to initiate the systematic build-up of such knowledge to support the right evacuation behaviour in the different kinds of evacuation scenarios that can occur.

Many evacuations from trains occur in situations without any obvious or immediate threat and can thus be organised by the staff, but even so the evacuations and the situations preceding the evacuation present risks. Since such situations occur quite frequently, they can if studied: firstly give important information on human safety behaviour in evacuation situations; and secondly, support identification of new risks which in turn must be managed.

3. Data Collection from Real Evacuation Situations

Data were collected from real evacuation situations by the use of questionnaires, answered by train drivers, ***dispatchers*** and passengers, over a three-year period. Some of the results from questionnaires filled out by passengers and train staff that had experienced a real evacuation situation are reported in this paper.

Data were collected from passengers and train staff from six Swedish train operating companies, but the majority of the questionnaires came from just three of these. Different questionnaires were used for passengers and staff. The group consisting of 160 train staff answered the questionnaires based on 113 different evacuation situations, so in some cases more than one person from the same evacuation situation answered the questionnaire. 51% of the questionnaires were answered by drivers, and the rest from other train staff.

125 questionnaires were collected from passengers from 33 different evacuation situations. Most questionnaires were collected from commuter passengers in the Stockholm area.

The questionnaires contained multiple choice as well as open questions. The questions concerned communication and information, time taken to decide on and conduct the evacuation, equipment to support evacuation, feelings about the evacuation, training and procedures as well as open questions on suggestions for improvements. Descriptive results are presented below as frequencies (see Figure 20-1).

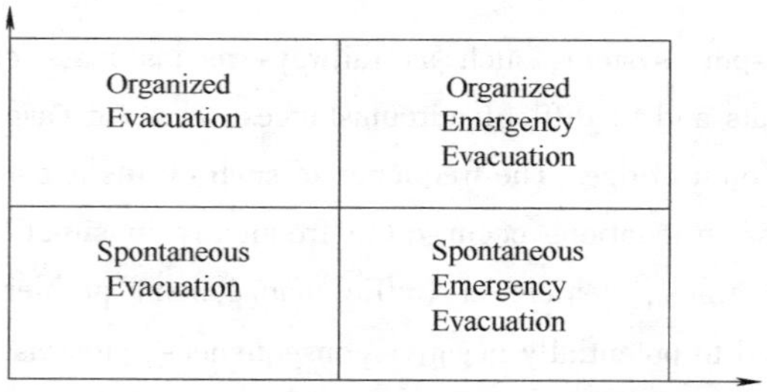

Organized Evacuation	Organized Emergency Evacuation
Spontaneous Evacuation	Spontaneous Emergency Evacuation

Figure 20-1 A Model For Train Evacuation. Evacuations Described from Two Dimensions: Level of Control and Level of Threat

4. Which Evacuation Situations Can Occur?

As presented above most evacuations occur in situations where there is no immediate threat. In railway settings it is necessary to organize the evacuation to be able to control the risks of electricity accidents and of being hit by another train. The need for organization of the evacuation is also stated in the railway authority's regulations. Different types of evacuation ***scenarios*** could be identified.

Two main ***dimensions*** were identified to describe the evacuation scenarios: level of threat and level of control (the ability to organize the evacuation).

Based on these two dimensions four types of evacuation scenarios are identified:

- organized evacuation (high level of control and low level of threat);
- organized emergency evacuation (high level of control and high level of threat);

- spontaneous evacuation (low level of control and low level of threat);
- spontaneous emergency evacuation (low level of control and high level of threat).

The data collected in this study were from situations with a low level of immediate threat such as fire. Even so there are other risks when people are outside the train. Also, new risks are introduced because other consequences develop over time due to difficult conditions inside the train, such as crowding, high temperatures, lack of ***fresh air***, heat, cold etc., while the passengers have to wait to be evacuated.

5. Causes for Evacuation and Physical Environment

The train staff answered a question about the causes of evacuation. 38% of the evacuations were due to vehicle problems, and 28% due to a broken ***aerial*** line. Smoke in the train caused 14% of the evacuations, hold-up in the traffic 8%, and collision with a person or animal 10%. Other events, for example fallen trees on the track, caused 11% of the evacuations.

71% of the evacuations occurred in a place where there was more than one track. The environmental and weather conditions varied between the evacuations. 35% of the evacuations were made in dark conditions. 15% of the evacuations were carried out in very cold weather, and 15% in warm weather. In some cases, the temperature was very high.

6. Experience and Training of the Staff

50% of train staff had more than ten years of experience in their profession, and 12% less than two years. The drivers had the longest experience. 40% of the staff had never been in an evacuation before. 27% had evacuated once before, and 33% had evacuated several times.

61% of the staff stated that they had both theoretical and practical education concerning evacuation, but 6% had neither education nor training. A majority emphasized the importance of training and education, especially the need for more practical training.

7. Communication with Professionals and Passengers

In the case of an evacuation decisions have to be made by the train staff and the train dispatching centre. For that reason the train staff and the train dispatching centre have to be able to communicate. In most of the evacuations the staff experienced communication with other train staff working well, but in some cases there were problems in the communication.

In 8% of the evacuations, the staff experienced problems in the communication with other train staff. In about 24% of the evacuations, there were problems in contacting the train dispatching centre. The causes were problems with telephones, radio and/or speaker systems.

According to the train staff, the evacuation in 38% of the cases was announced on the loud speaker system on the train. Another way of communicating the evacuation to the passengers was that the train staff announced the evacuation verbally in the carriages. In 6% of the evacuations, the evacuation was not announced at all. In 22% of the evacuations, the train staff experienced problems in communication with passengers. In 84% of the cases information was given only in Swedish.

66% of the passengers stated that they received the information about evacuation on the loud speakers, while 25% of the passengers received the information from the staff on the train. 10% of the passengers got the information from fellow-passengers, in some other way, or didn't get it at all. Many of the passengers emphasized in the questionnaires the importance of information, and pointed out, among other things, problems such as lack of information, unclear information and that there were no staff easy to approach.

8. Time Spent Waiting for Evacuation

The train staff was asked to estimate the time from the train stopped until the evacuation started. In 16% of the cases the time was estimated to less than 10 minutes, and in 57% of the cases to more than 30 minutes. When the evacuation was fast this was due to for example smoke in the train. The cases where the time until the evacuation started was longer than 30 minutes, the reason for evacuation was in most cases vehicle problems or a broken aerial line. In some extreme cases the time from the train stopped until the start of the evacuation have been three to four hours.

Several of the train staff stated in the questionnaire that the time waiting on the train often was too long, frequently because the decision about evacuation took too long or it took a long time for specially trained staff who take care of broken aerial lines to get to the train. If the train staff and the passengers had to wait on the train for a long time the risk of spontaneous evacuation greatly increased. A member of train staff commented:

"Thank God it was 6 o'clock in the morning. If it would have been later in the day, it would have been impossible to keep 200 passengers in the train the whole time."

Many passengers emphasized that the evacuation took a long time. For some of the passengers the conditions were also difficult, which made the waiting unpleasant. In some cases, the train was standing still for hours in winter cold, which caused a very low temperature inside the train when the power supply ***ran out***. In other cases very high summer temperatures caused heat inside the train which almost caused people to faint. Passengers that were stuck in the train for hours reported about the need to get water to drink and to have access to a toilet.

Some participants in this study described the situation as:

"It was terribly hot on the train, for a while I thought I was going to faint." "It was extremely hot in our carriage since it was very crowded and people everywhere. The situation could have been improved if the firemen, as they passed through the carriage after we had waited for an hour, had broken a few windows. This would have given us fresh air. One and a half hours of waiting standing up in a tilting train is more than most people can handle. Fresh air would have made it easier."

New Words and Expressions

evacuation [ɪˌvækjʊ'eɪʃ(ə)n] n. 撤退，抽空，排泄，腾出
dispatcher [dɪs'pætʃə] n. （急件等的）发送人
scenario [sɪ'naːrɪəʊ] n. 方案，（可能发生的）情况，前景，［戏，影视］脚本
dimension [dɪ'menʃən] n. 尺寸，规模，面积，容积
fresh air 新鲜空气
aerial ['eərɪəl] n. 天线
ran out 耗尽，抛弃，放弃

Exercises

1. What's the problem for evacuation from trains? What about the research on evacuation from trains?
2. What should we study during research on evacuation from trains?
3. Which evacuation situations can occur?
4. What can you know from the questionnaires?
5. How did participants in this study describe the situation for the time spent waiting for evacuation?

科技英语的常用句型

在科技英语的写作中，要求文句要科学严谨，通俗易懂，精炼扼要，在写作方式和句型、用词上有较高的要求。科技英语的写作往往会用到一些常用的表达方式，如描述实验原理、实验现象，或者摘要、引言、结论的书写，都会用到一些常用的句型，下面做了简单总结。

1. 摘要

（1）引言

1）引入。

We review evidence for...

This paper outlines some of the basic methods and strategies and discusses...

We present/describe the background of...

We summarize searches for...

2）写作目的。

The purpose of this paper is...

The primary goal of this research is...

Our goal has been to provide...

We made these experiments in order to show a correlation...

The study of... was made in order to clarify...

The evaluate and compare the different degree of...

To evaluate and compare the different degree of...

3）研究范围。

The article contains some practical recommendation on...

The scope of the research covers...

Subjects covered includes...

The paper contains the specific topics on...

4）研究重点。

This report concentrates on...

Attentions is concentrated on...

Particular attention is/was paid to...

There has been a focus on... and attention is being paid to...

(2) 方法

1) 介绍研究和实验过程。

This paper examines how...

We present an analysis of...

The paper analyzes the possibility of...

The study identifies some procedures for...

The article discusses the method of calculation of...

2) 介绍研究和试验的方法。

The experiments on... have been carried out using...

The tests were carried out on...

This study presents estimates of...

3) 介绍应用。

Such a statistical method has been applied to...

The application of... Is to develop and maintain...

As an application, we implement...

Using... to monitor expression in these vectors...

(3) 结果

The result of observation show that...

The findings of our research on... is...

On the basis of..., the following conclusions can be drawn...

In conclusion, the results show that...

(4) 结论

We thus conclude that...

Exciting new research has elaborated several important and unexpected findings that explain...

This paper is mainly devoted to...

We showed that it is possible to identify...

A simplified diagram of... and... is taken as reference.

Recommendations are made regarding...

Our work involving studies of... did not prove to be encouraging.

The theory by... seemed quite competent.

The development of tools to... is desirable during...

There is a growing need for/demand for... By the increasing incidence of...

2. 引言

(1) 研究目的

1) 论文导向。

This purpose of this paper is...

The aim of this report is...

The object of the paper is to...

2) 研究导向。

The purpose of the experiment reported here was...

The aim of this study was...

The objective of this research was to determine. . .

In this research described here, . . .

In this study. . .

In the present investigation, . . .

In this research, . . .

（2）研究价值

Although much research has been devoted to A, . . .

While much work has been done on A, . . .

Although many studies has been published concerning A, . . .

While many researchers have investigated A, . . .

Although much literature is available on A, little research has been done on B.

. . . little attention has been paid to B.

. . . little information is available on B.

. . . little work has been published on B.

. . . few researchers

（3）背景和历史

It seems (exceedingly, enormously) difficulty to obtain knowledge of the problem. . .

It proves (totally, quite) impossible to try. . .

It is rather difficult to solve the problem. . .

The problem involves (certain, tremendous) difficulties.

It is no easy to present (reveal, analyze, discuss) the problem in all its complexity (in every detail).

（4）问题的范围

The main aspect (core, essence) of problem is. . .

Studies of these effects cover various aspects of. . .

Our studies with this technique confine to. . .

The problem is within the scope of. . .

Our problem lies beyond the range of. . .

3. 正文

（1）理论说明

Our theory is based on the assumption that. . .

This theory proceeds from the idea (principle) of. . .

The underlying concept of the theory is as follows.

There is a similar (alternative, tentative) theory that. . .

The basic (essential, fundamental) feature of this theory is. . .

The object of this theory is to. . .

The newly advanced theory has some advantages (assets, strong point, positive features, deficiencies, drawback, inadequacies, flaws, shortcomings). . .

The validity of the theory has become obvious in the light of recent findings.

This newly-developed theory finds experimental support. . .

The theory received universal recognition (general acceptance).

(2) 公式推导

. . . is given by:

. . . as follows:

. . . as in the following:

. . . the following equation is obtained:

. . . can be expressed as/derived/written/described by/represented as. . .

(3) 方法介绍

The method of. . . was first developed by. . .

The method of. . . came into use as long ago as. . .

The original proposal of this method was first published in. . .

The method we used differs from the conventional one.

This method allows us to demonstrate. . .

This method is capable of providing. . .

(4) 实验描述

We made this experiment to show a correlation between. . .

We carried out this experiment to demonstrate a certain phenomena. . .

We performed this experiment to elucidate the mechanism of. . .

We initiated this experiment to evaluate the hypothesis of. . .

Experiment on. . . are made to. . .

The experiments reported here demonstrate a variety of changes in. . .

Our experiments with. . . furnish some new information of. . .

Further experiments in this area leads to conclude that. . .

From these experiment we can conclude that. . .

4. 结论

(1) 结果的意义

The results presented in this paper are. . .

The above findings can be viewed as follows. . .

We can consider these results as fully reliable. . .

The results reported here prove the hypothesis that. . .

(2) 导致结论

Our findings suggests that. . .

These findings lead the author to a conclusion that. . .

The research work has brought about a discovery of. . .

Further progress can be providing by this experiment.

In the future, we will extend the present studies to. . .

Human Factors in the Railway System Safety Analysis Process

In continental Europe, the consideration of human factors does not have a long tradition. Although a high percentage of accidents are accounted to human error, the integration of human contribution into the system's safety is often rudimentarily analyzed in railway engineering. Unfortunately, outdated approaches of continuous automation or usage of fixed human error probabilities for ***quantitative risk analyses*** can still be found. Therefore, this article takes a point of view quite close to railway engineering practice and tries to give some answers on the need for human factors integration.

1. *Current Situation for Human Factors and Railway System Safety*

(1) Human Factors in the Railway System's Life-Cycle

The focus of this paper is on dependability, understood as reliability, availability and maintainability and particularly safety. The most important and well-known standard for the system design of railway technical components EN requires integration of human factors. In spite of emphasizing the importance of ***consideration*** of human factors of railway system's staff, the standard provides ***sparse*** information on the way of integration.

The workplace of humans at the sharp end, i. e. the railway system in operation, and corresponding safety-related risks is shaped by design engineers, the employer and the operator himself. Hereby, the designer is not necessarily only located at the manufacturer as the operating company often sets the requirements very closely, in the railway industry.

If ***safety assessments*** consider human factors, the scope is frequently limited to the designer's evaluation of the risk that evolves from the operator's task. Thereby firstly, the designer neither estimates divergences to the user's risk perception in the moment of operation nor takes his own error into account. Though, systematic errors are to be controlled by requirements determined by safety ***integrity*** levels and new safety management systems. The design engineer has an external and stable perspective to evaluate risk for a human-machine-system, though the operator has a very flexible point of view for his risk control (Vanderhaegen, 2004). Secondly, the employer and his risk assessment and safety-related measures are cut out of these considerations. We argue that future risk assessments have to integrate not only the human factors of the operator, but also the employer's measures and the risk of ***human error*** in the design phase.

In the remainder of this paper, we focus on the design for human operators in the operation and maintenance phase due to the long duration and the comparatively high risks this phase involves. In addition, we limit the scope to train drivers—meaning staff of the railway operating company, undertaking the task of driving trains—and signallers—meaning staff of the infrastructure managing company undertaking the task of authorizing the movement of trains (definitions taken from the Technical Specifications for Interoperability). These two work stations influence railway performance and safety directly and on-line in the moment of operation.

(2) Quantitative Risk Analysis

The European standard EN 50129 (CENELEC, 2003) requires a quantitative risk analysis for safety-related railway systems. Hazards are to be identified and the corresponding risk is estimated. If the risk is not negligible, it has to be proven that the final product fulfils the safety requirements. The actual hazard rate of a safety-related function must not be higher than the associated tolerable hazard rate. The reliability of functions implemented by technical components can in most cases be estimated sufficiently. The human reliability is usually characterized by an estimation of a human error probability. Bringing together technical reliability and human performance is vital for railway transport because most safety-related functions are implemented by a combination of technical systems and human actions.

2. Discussion of Existing Approaches

(1) Human Error in Quantitative Risk Analysis

When pursuing a classic risk assessment, the railway practitioner has difficulties in integrating the human reliability into the analysis methods: This is due to the lack of valid data for human error probabilities in the railway domain. In German railway engineering practice, sometimes the constant human error probability of 10–3 is chosen in spite of the high variability of human performance. More sophisticated risk analyses refer to the values for human error in railway transport published by Hinzen (1993). 18 fixed probabilities are presented in dependence of stress level, surrounding conditions and Rasmussen's (1983) three levels of behaviour: knowledge-based, rule-based and skill-based. Feldmann et al. (2008) show that neither Hinzen's values are fully proven to be valid for the railway transport, nor do accident statistics provide comprehensive data to derive human error probabilities. In order to obtain probabilities for certain working conditions, the classic approach is to use Human Reliability Assessment (HRA) methods.

(2) Existing Techniques for Human Reliability Assessment

The steps of the most HRA approaches are to perform a qualitative analysis of the task by conducting a task analysis together with an assessment of possible human errors. Subsequently, Human Error Quantification (HEQ) methods can be applied.

The technique Analysis of Consequences of Human Unreliability (ACIH) (Vanderhaegen, 2001) represents a non-probabilistic technique for human reliability assessment. The non-quantitative method can be used when a qualitative integration of human error into risk assessment is sufficient. One of the most common HEQ methods is called THERP (Technique of Human Error Rate Prediction). First, the task under consideration is ***decomposed*** into several individual tasks. The corresponding nominal probability from the handbook is to be adjusted by a set of so called Performance Shaping Factors (PSF) and the calculation by prescribed rules. The nominal probabilities in THERP were recorded in the nuclear power industry. Not only does railway transport differ from that application, also the values reflect a technology and control system design that is actually no more applicable. THERP was already initially applied to the railway driving task (Chaali-Djelassi et al., 2007), but without the adaptation of the probabilities by influencing factors. Another technique, HEART (Human Error Assessment and Reduction Technique; Williams, 1986) is also based on nominal probabilities that are adjusted by so called error producing conditions. To adapt this technique to railway work is challenging and possibly exposed to discussions due to the original estimations of the probabilities with different workplaces in mind.

Some of the latter drawbacks of existing techniques were motivation for the British Rail Safety and Standards Board to develop a first railway-specific HRA method. A review exposed the high ***complexity*** and some disadvantages of the new technique. Analyzing the 29 factors that influence the human performance, a high degree of overlapping and interdependence between the PSF can be observed. Consider for example the set of factors unfamiliarity, driver experience and technique learning. In rail-HEQ—like in several HRA methods—performance shaping factors represent a complex set and are neither well-separated nor visualized. PSF like concentration and fatigue can themselves be influenced by other factors before. Here, we propose a ***clarification***, i. e. a differentiation between cause and effects, in order to avoid faulty double-representation of influence factors.

At least since the human error taxonomy by Reason (1990) it has been known that human malfunctions appear in several shapes. Some classic risk analysis techniques are capable of modelling errors of omission and errors of commission; i. e. slips or lapses. Three other error types make predictive human error more difficult to and not yet fully covered by analysis techniques: intentional human errors, unexpected human actions, and human recovery of errors.

(3) Human Variability and Resilience

The drawbacks of existing approaches and the great variability of human error have led to new ideas. Understanding human variability as a capability and not as a threat, understanding safety as the presence of adaptability instead of the absence of weaknesses or human errors has become known as resilience engineering. See Quéva (2008) for an approach of adaptability and reactivity to describe human variability when driving an urban train. The accident analysis method FRAM (Hollnagel, 2004) has been applied to the railway domain (Belmonte, 2009). Needless to say that, due to their distance from the classical analysis of human error, these approaches do not respond to the need of the CENELEC railway standards.

(4) Discussion and Approach

Summing up the considerations of the last three subsections, neither fixed values for human error probabilities nor one of the existing HRA techniques can fully meet the requirements of the rail standards. There is a demand for the development of a serviceable tool applicable in railway engineering practice, with components less fine-grained to stay practicable whilst providing at least semiquantitative assessments of human error and performance. Without limiting the scope to non-intentional errors, the paper proposes a clarification and railway applicable structuring of performance shaping factors with a work system model.

3. Human-Barrier-Interaction in Railway Systems

In order to study organizational and physical influence factors in a better way and to take consequences of human performance into account, the utilisation of safety barriers is proposed as another approach to human error.

Barriers represent safety mechanisms that are installed to prevent undesired events from taking place or to protect against its consequences. The most common taxonomy of barriers distinguishes between material barriers being physically in the system, functional barriers creating dependencies, symbolic and immaterial barriers, the latter both being for example signs or rules. For the classification of safety barriers in terms of a process model, a three step structure was proposed: Barriers of prevention that prevent an undesired initial event from taking place, barriers of correction that recover the situation and barriers of containment that lessen the severity of the consequences.

While a physical and a functional barrier system executes the barrier function itself, symbolic and immaterial barriers request an action and its performance represents the barrier function. Due to the dependency on human actions, these barrier systems can generally be estimated as being medium or less efficient.

The efficiency of barriers in terms of human-barrier-interaction should be analyzed in a more profound manner. Thereby, the approach should not be limited to ***violations*** but to human performance in general. The following barrier properties are proposed as exemplary detail criteria for a well-functioning of human-barrier interaction: ability, benefit and ease to deactivate; temporal presence, perceptibility of the barrier and its status and finally temporal and ***spatial*** distance between barrier system and barrier function.

In order to apply this idea to the railway system, there is a necessity for barrier identification. Up to now, two initial approaches can be found in literature. Consider the protection against overspeed in the train-driving task as an example. A symbolic barrier of prevention is the speed indicator which is continuously visible to the driver. The high perceptibility certainly has a positive effect on the probability of well-functioning of this particular human-barrier system. In contrary, an advance speed limit sign appears only punctually and involves a certain delay to execution of the barrier function. The train control system (preventing overspeed in a technical way) is a functional barrier of correction that can be deactivated under circumstances and whose state is disadvantageously not always well perceivable for the operator.

Regard the link to performance shaping factors: While the design of the speed indicator instrument is a physical factor (design of human-machine-interface), the actual information represents a dynamically changing input. Rules against overspeed constitute immaterial barriers of prevention and are organizational factors.

So, the approach of human-barrier-interaction supports the analysis of performance shaping factors and their influence degree. Furthermore, barriers also permit the study of consequences of malfunctioning as barriers can overlap or secure each other. Last, the analysis of barrier regimes also gives hints for redesign and error reduction.

New Words and Expressions

quantitative risk analysis　定量风险分析
consideration [kənsɪdə'reɪʃ(ə)n]　adj. 顾及，报酬，斟酌，仔细考虑
sparse [spɑːs]　adj. 稀少的，稀疏的，零落的
safety assessment　安全评估，安全性评价，安全研究
integrity [ɪn'tegrɪti]　n. 完整，诚实正直，完好
human error　人为误差
decompose [ˌdiːkəm'pəʊz]　v. 分解，风化，(使)还原，(使)腐烂
complexity [kəm'pleksəti]　n. 复杂性，难懂，难题，难以理解的局势
clarification [ˌklærɪfɪ'keɪʃ(ə)n]　n. 说明，净化，澄清(作用)，澄清法
violation [vaɪə'leɪʃn]　n. 侵犯，违背，妨害，违例
spatial ['speɪʃ(ə)l]　adj. 空间的

附录　机构名称一览表

机构名称	英文全称
国际组织	
国际劳工组织	International Labor Organization
世界卫生组织	World Health Organization
国际职业安全健康信息中心	International Occupational Safety and Health Information Centre
国际劳动监督协会	International Association of Labour Inspection
国际职业卫生学会	International Commission on Occupational Health
国际社会保障协会	International Social Security Association
职业安全健康网	Occupational Health and Safety Resources Net
欧洲职业安全健康局	European Agency for Safety and Health at Work
加拿大-欧洲工作地安全和健康联盟	Canada and European Union Cooperation on Workplace Safety and Health
亚太职业安全健康组织	Asia Pacific Occupational Safety and Health Organizations
欧洲过程安全中心	European Process Safety Centre
美国	
美国职业安全健康局	Occupational Safety & Health Administration
美国职业安全健康复审委员会	Occupational Safety and Health Review Commission
美国国家安全委员会	National Safety Council
美国矿山安全健康局	Mine Safety and Health Administration
美国联邦职业安全健康项目	Federal Employee Occupational Safety and Health
美国环境健康科学院	National Institute of Environmental Health Sciences
美国职业安全健康研究所	National Institute for Occupational Safety and Health
国家运输安全委员会	NTSB-National Trans-Potation Safety Board
美国化工安全与危害调查局	U. S. Chemical Safety and Hazard Investigation Board
美国联邦应急管理局	Federal Emergency Management Agency
加拿大	
加拿大职业安全健康网	Canada’s National Occupational Health and Safety Web Site
加拿大工业事故预防协会	Industrial Accident Prevention Association
加拿大职业安全健康中心	Canadian Centre for Occupation Health and Safety
加拿大安大略建筑安全协会	Construction Safety Association of Ontario，Canada
英国	
英国职业安全健康学院	Institution of Occupational Safety and Health
英国健康安全局	Health and Safety Executive，UK
英国健康安全委员会	Health and Safety Commission，UK
英国皇家事故预防学会	Royal Society for the Prevention of Accidents，UK

（续）

机构名称	英文全称
其他国家	
日本工业健康学院	National Institute of Industrial Health, Japan
日本煤炭能源中心	Japan Coal Energy Center
日本工业安全健康协会	Japan Industrial Safety and Health Association
日本国际职业安全健康中心	Japan International Center for Occupation Safety and Health
新加坡工业安全部	The Department of Industrial Safety, Singapore
德国联邦职业安全健康学院	Federal Institute for Occupational Safety and Health
爱尔兰健康和安全署	Health and Safety Authority Ireland
南非职业安全健康学会	Association of Societies for Occupational Safety and Health
澳大利亚职业安全健康委员会	National Occupational Health and Safety Commission
澳大利亚安全委员会	National Safety Council of Australia
新西兰职业健康安全网	Occupational Safety and Health Service of the Department of Labour
马来西亚职业安全健康学院	National Institute of Occupational Safety and Health
菲律宾职业安全健康中心	Occupational Safety and Health Center
埃及职业安全健康研究所	National Institute of Occupational Safety & Health
印度劳工部职业安全健康处	The Indiana Occupational Safety and Health Division
韩国职业安全健康局	Korea Occupational Safety and Health Agency
中国	
国家安全生产监督管理总局	State Administration of Work Safety
国家煤矿安全监察局	State Administration of Coal Mine Safety
国家安全生产应急救援指挥中心	National Administration for Work Safety Emergency Response
国家食品药品监督管理总局	China Food and Drug Administration
国家食品安全风险评估中心	China National Center for Food Safety Risk Assessment
中国职业安全健康协会	China Occupational Safety and Health Association
中国安全生产协会	China Association of Work Safety
中国化学品安全协会	China Chemical Safety Association

参考文献

[1] Gitelman V, Hendel L, Carmel R, et al. An Examination of the National Road-Safety Programs in the Ten World's Leading Countries in Road Safety [J]. European Transport Research Review, 2012, 4 (4): 175-188.

[2] Limbri H, Gunawan C, Rosche B, et al. Challenges to Developing Methane Biofiltration for Coal Mine Ventilation Air: A Review [J]. Water, Air, & Soil Pollution, 2013, 224 (6).

[3] Levitt M. Concrete Materials Problems and Solutions [M]. Boca Raton: Spon Press, 1997.

[4] Hinze J. Construction Safety [J]. Safety Science, 2008, 46 (4): 565.

[5] Huang L, Liang D. Development of Safety Regulation and Management System in Energy Industry of China: Comparative and Case Study Perspectives [J]. Procedia Engineering, 2013, 52: 165-170.

[6] Moridi M A, Kawamura Y, Sharifzadeh M, et al. Development of Underground Mine Monitoring and Communication System Integrated ZigBee and GIS [J]. International Journal of Mining Science and Technology, 2015, 25 (5): 811-818.

[7] Taylor G, Easter K, Hegney R. Enhancing Occupational Safety and Health [M]. New York: Elsevier, 2004.

[8] Kecklund L, IngridAnderzén, Petterson S, et al. Evacuation from Trains—The Railway Safety Challenge [J]. Rail Human Factors around the World, 2012: 815-823.

[9] Erdogan S. Explorative Spatial Analysis of Traffic Accident Statistics and Road Mortality among the Provinces of Turkey [J]. Journal of Safety Research, 2009, 40 (5): 341-351.

[10] Naghizadeh A, Mahvi A H, Jabbari H, et al. Exposure Assessment to Dust and Free Silica for Workers of Sangan Iron Ore Mine in Khaf, Iran [J]. Bulletin of Environmental Contamination and Toxicology, 2011, 87 (5): 531-538.

[11] Layden W M. Food Safety [J]. Society, 1994, 31 (4): 20-26.

[12] McElhatton A, Marshall R J. Food Safety a Practical and Case Study Approach [M]. New York: Springer, 2007.

[13] Day S J, Carras J N, Fry R, et al. Greenhouse Gas Emissions from Australian Open-Cut Coal Mines: Contribution from Spontaneous Combustion and Low-Temperature Oxidation [J]. Environmental Monitoring and Assessment, 2010, 166 (1-4): 529-541.

[14] Dionne G. Handbook of Insurance [M]. New York: Springer, 2013.

[15] Eidson J V, Reese C D. Handbook of OSHA Construction Safety and Health [M]. Boca Raton: CRC Press, 2006.

[16] Bahadori A. Hazardous Area Classification in Petroleum and Chemical Plants [M]. Boca Raton: CRC Press, 2013.

[17] Tweedy J T. Healthcare Safety Management [M]. Boca Raton: CRC Press, 2005.

[18] Novak B. Human Factors Contributions to Consumer Product Safety [M]. London: Taylor & Francis, 2014.

[19] Hammerl M, Vanderhaegen F. Human Factors in the Railway System Safety Analysis Process [M]. London: Taylor & Francis, 2012.

[20] Tatiya R R. Industrial Safety [M]. Boca Raton: CRC Press, 2010.

[21] Mu L, Ji Y. Integrated Coal Mine Safety Monitoring System [M]. Berlin, Heidelberg: Springer Berlin Heidelberg, 2012.

[22] Khan F, Rathnayaka S, Ahmed S. Methods and Models in Process Safety and Risk Management: Past, Present and Future [J]. Process Safety and Environmental Protection, 2015, 98: 116-147.

[23] O'Reilly M. Occupational Ergonomics [M]. Boca Raton: CRC Press, 2012.

[24] Perdikaris J. Physical Security and Environmental Protection [M]. Boca Raton: CRC Press, 2014.

[25] Murthy, Prabhakar D N, Rausand, et al. Product Reliability [M]. London: Springer, 2008.

[26] Ray S K, Singh R P. Recent Developments and Practices to Control Fire in Undergound Coal Mines [J]. Fire Technology, 2007, 43 (4): 285-300.

[27] Resources Safety D O C A. Refuge Chambers in Underground Metalliferous Mines [S]. Department of Industry and Resources, 2008.

[28] Raheja D. Risk Assessment and Risk Management [M]. Boca Raton: CRC Press, 2014.

[29] Szulik A, Lebecki K, Cybulski K. Risk of Coal Dust Explosion and its Elimination [M]. London: Taylor & Francis, 2004.

[30] Uno N, Oshima Y, Thompson R. Road Safety [M]. Boca Raton: CRC Press, 2013.

[31] Hudson J A. Rock Failure Mechanisms Illustrated and Explained [M]. Boca Raton: CRC Press, 2010.

[32] Huang Y, Chen P Y, Grosch J W. Safety Climate: New Developments in Conceptualization, Theory, and Research [J]. Accident Analysis & Prevention, 2010, 42 (5): 1421-1422.

[33] Yost C P. Safety Education [M]. New York: Educational Research Association, 1962.

[34] Bhise V D. Safety Engineering in Product Design [M]. Boca Raton: CRC Press, 2013.

[35] Yates W D. Safety Professional's Reference and Study Guide [M]. Boca Raton: CRC Press, 2010.

[36] Tian J, Chen H. Special Equipment Safety Analysis and Countermeasures [C]. International Conference on Uncertainty Reasoning and Knowledge Engineering. IEEE, 2011: 161-164.

[37] Pouliakas K, Theodossiou I. The Economics of Health and Safety at Work: An Interdiciplinary Review of the Theory and Policy [J]. Journal of Economic Surveys, 2013, 27 (1): 167-208.

[38] Wallace K, Prosser B, Stinnette J D. The Practice of Mine Ventilation Engineering [J]. International Journal of Mining Science and Technology, 2015, 25 (2): 165-169.

[39] Wiley H N. Universal Access in the Information Society [M]. Berlin Heidelberg: Springer, 2005.

[40] Shin I, Oh J, Yi K H. Workers' Compensation Insurance and Occupational Injuries [J]. Safety and Health at Work, 2011, 2 (2): 148-157.

[41] 孙昌坤．实用科技英语翻译 [M]．北京：对外经济贸易大学出版社，2013.

[42] 崔小清．科技英语翻译与写作 [M]．西安：西北工业大学出版社，2011.

[43] 谢国章．科技英语翻译技巧 [M]．北京：人民邮电出版社，1987.

[44] 田文杰．科技英语教程 [M]．西安：西安交通大学出版社，2008.

[45] 王泉水．科技英语翻译技巧 [M]．天津：天津科学技术出版社，1991.

[46] 西北工业大学外语教研室．科技英语翻译初步 [M]．北京：商务印书馆，1979.

[47] 黄荣恩．科技英语翻译浅说 [M]．北京：中国对外翻译出版公司，1981.

[48] 王册．科技英语翻译技巧 [M]．哈尔滨：黑龙江科学技术出版社，1985.

[49] 杨跃，马刚．实用科技英语翻译研究 [M]．西安：西安交通大学出版社，2008.

[50] 王纯真．科技英语疑难句型的理解与翻译 [M]．南京：东南大学出版社，2013.

[51] 夏喜玲．科技英语翻译技法 [M]．郑州：河南人民出版社，2007.

[52] 郑仰成，赵萱．科技英语翻译教程 [M]．北京：外语教学与研究出版社，2006.